Newtonian Physics

The **Light and Matter** series of
introductory physics textbooks:

Newtonian Physics

Benjamin Crowell

www.lightandmatter.com

Light and Matter

Fullerton, California
www.lightandmatter.com

Edition 2.2
rev. 2003-03-28

ISBN 0-9704670-1-X

To Paul Herrschaft and Rich Muller.

Brief Contents

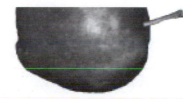

Contents

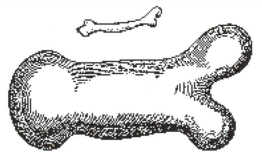

Motion in One Dimension   53

Motion in Three Dimensions 139

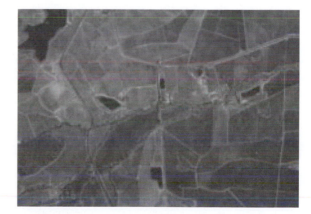

Preface

Why a New Physics Textbook?

We assume that our economic system will always scamper to provide us with the products we want. Special orders don't upset us! I want my MTV! The truth is more complicated, especially in our education system, which is paid for by the students but controlled by the professoriate. Witness the perverse success of the bloated science textbook. The newspapers continue to compare our system unfavorably to Japanese and European education, where depth is emphasized over breadth, but we can't seem to create a physics textbook that covers a manageable number of topics for a one-year course and gives honest explanations of everything it touches on.

The publishers try to please everybody by including every imaginable topic in the book, but end up pleasing nobody. There is wide agreement among physics teachers that the traditional one-year introductory textbooks cannot in fact be taught in one year. One cannot surgically remove enough material and still gracefully navigate the rest of one of these kitchen-sink textbooks. What is far worse is that the books are so crammed with topics that nearly all the explanation is cut out in order to keep the page count below 1100. Vital concepts like energy are introduced abruptly with an equation, like a first-date kiss that comes before "hello."

The movement to reform physics texts is steaming ahead, but despite excellent books such as Hewitt's **Conceptual Physics** for non-science majors and Knight's **Physics: A Contemporary Perspective** for students who know calculus, there has been a gap in physics books for life-science majors who haven't learned calculus or are learning it concurrently with physics. This book is meant to fill that gap.

Learning to Hate Physics?

When you read a mystery novel, you know in advance what structure to expect: a crime, some detective work, and finally the unmasking of the evildoer. When Charlie Parker plays a blues, your ear expects to hear certain landmarks of the form regardless of how wild some of his notes are. Surveys of physics students usually show that they have *worse* attitudes about the subject after instruction than before, and their comments often boil down to a complaint that the person who strung the topics together had not learned what Agatha Christie and Charlie Parker knew intuitively about form and structure: students become bored and demoralized because the "march through the topics" lacks a coherent story line. You are reading the first volume of the **Light and Matter** series of introductory physics textbooks, and as implied by its title, the story line of the series is built around light and matter: how they behave, how they are different from each other, and, at the end of the story, how they turn out to be similar in some very bizarre ways. Here is a guide to the structure of the one-year course presented in this series:

1 Newtonian Physics   *Matter* moves at constant speed in a straight line unless a force acts on it. (This seems intuitively wrong only because we tend to forget the role of friction forces.) Material objects can exert forces on each other, each changing the other's motion. A more massive object changes its motion more slowly in response to a given force.

2 Conservation Laws   Newton's matter-and-forces picture of the universe is fine as far as it goes, but it doesn't apply to *light*, which is a form of pure energy without mass. A more powerful world-view, applying equally well to both light and matter, is provided by the conservation laws, for instance the law of conservation of energy, which states that energy can never be destroyed or created but only changed from one form into another.

3 Vibrations and Waves   *Light* is a wave. We learn how waves travel through space, pass through each other, speed up, slow down, and are reflected.

4 Electricity and Magnetism   *Matter* is made out of particles such as electrons and protons, which are held together by electrical forces. *Light* is a wave that is made out of patterns of electric and magnetic force.

5 Optics   Devices such as eyeglasses and searchlights use *matter* (lenses and mirrors) to manipulate *light*.

6 The Modern Revolution in Physics   Until the twentieth century, physicists thought that *matter* was made out of particles and *light* was purely a wave phenomenon. We now know that both light and matter are made of building blocks that have both particle and wave properties. In the process of understanding this apparent contradiction, we find that the universe is a much stranger place than Newton had ever imagined, and also learn the basis for such devices as lasers and computer chips.

A Note to the Student Taking Calculus Concurrently

Learning calculus and physics concurrently is an excellent idea — it's not a coincidence that the inventor of calculus, Isaac Newton, also discovered the laws of motion! If you are worried about taking these two demanding courses at the same time, let me reassure you. I think you will find that physics helps you with calculus while calculus deepens and enhances your experience of physics. This book is designed to be used in either an algebra-based physics course or a calculus-based physics course that has calculus as a corequisite. This note is addressed to students in the latter type of course.

It has been said that critics discuss art with each other, but artists talk about brushes. Art needs both a "why" and a "how," concepts as well as technique. Just as it is easier to enjoy an oil painting than to produce one, it is easier to understand the concepts of calculus than to learn the techniques of calculus. This book will generally teach you the *concepts* of calculus a few weeks before you learn them in your math class, but it does not discuss the *techniques* of calculus at all. There will thus be a delay of a few weeks between the time when a calculus application is first pointed out in this book and the first occurrence of a homework problem that requires the relevant technique. The following outline shows a typical first-semester calculus curriculum side-by-side with the list of topics covered in this book, to give you a rough idea of what calculus your physics instructor might expect you to know at a given point in the semester. The sequence of the calculus topics is the one followed by **Calculus of a Single Variable**, 2nd ed., by Swokowski, Olinick, and Pence.

chapters of this book	topics typically covered at the same point in a calculus course
0-1 introduction	review
2-3 velocity and acceleration	limits
4-5 Newton's laws	the derivative concept
6-8 motion in 3 dimensions	techniques for finding derivatives; derivatives of trigonometric functions
9 circular motion	the chain rule
10 gravity	local maxima and minima

chapters of
Conservation Laws

1-3 energy	concavity and the second derivative
4 momentum	
5 angular momentum	the indefinite integral

chapters of
Vibrations and Waves

1 vibrations	the definite integral
2-3 waves	the fundamental theorem of calculus

The Mars Climate Orbiter is prepared for its mission. The laws of physics are the same everywhere, even on Mars, so the probe could be designed based on the laws of physics as discovered on earth.

There is unfortunately another reason why this spacecraft is relevant to the topics of this chapter: it was destroyed attempting to enter Mars' atmosphere because engineers at Lockheed Martin forgot to convert data on engine thrusts from pounds into the metric unit of force (newtons) before giving the information to NASA. Conversions are important!

0   Introduction and Review

If you drop your shoe and a coin side by side, they hit the ground at the same time. Why doesn't the shoe get there first, since gravity is pulling harder on it? How does the lens of your eye work, and why do your eye's muscles need to squash its lens into different shapes in order to focus on objects nearby or far away? These are the kinds of questions that physics tries to answer about the behavior of light and matter, the two things that the universe is made of.

0.1   The Scientific Method

Until very recently in history, no progress was made in answering questions like these. Worse than that, the *wrong* answers written by thinkers like the ancient Greek physicist Aristotle were accepted without question for thousands of years. Why is it that scientific knowledge has progressed more since the Renaissance than it had in all the preceding millennia since the beginning of recorded history? Undoubtedly the industrial revolution is part of the answer. Building its centerpiece, the steam engine, required improved techniques for precise construction and measurement. (Early on, it was considered a major advance when English machine shops learned to build pistons and cylinders that fit together with a gap narrower than the thickness of a penny.) But even before the industrial revolution, the pace of discovery had picked up, mainly because of the introduction of the modern scientific method. Although it evolved over time, most scientists today would agree on something like the following list of the basic principles of the scientific method:

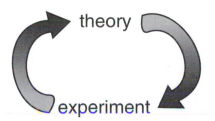

(1)*Science is a cycle of theory and experiment.* Scientific theories are created to explain the results of experiments that were created under certain conditions. A successful theory will also make new predictions about new experiments under new conditions. Eventually, though, it always seems to happen that a new experiment comes along, showing that under certain conditions the theory is not a good approximation or is not valid at all. The ball is then back in the theorists' court. If an experiment disagrees with the current theory, the theory has to be changed, not the experiment.

(2) *Theories should both predict and explain.* The requirement of predictive power means that a theory is only meaningful if it predicts something that can be checked against experimental measurements that the theorist did not already have at hand. That is, a theory should be testable. Explanatory value means that many phenomena should be accounted for with few basic principles. If you answer every "why" question with "because that's the way it is," then your theory has no explanatory value. Collecting lots of data without being able to find any basic underlying principles is not science.

(3) *Experiments should be reproducible.* An experiment should be treated with suspicion if it only works for one person, or only in one part of the world. Anyone with the necessary skills and equipment should be able to get the same results from the same experiment. This implies that science transcends national and ethnic boundaries; you can be sure that nobody is doing actual science who claims that their work is "Aryan, not Jewish," "Marxist, not bourgeois," or "Christian, not atheistic." An experiment cannot be reproduced if it is secret, so science is necessarily a public enterprise.

As an example of the cycle of theory and experiment, a vital step toward modern chemistry was the experimental observation that the chemical elements could not be transformed into each other, e.g. lead could not be turned into gold. This led to the theory that chemical reactions consisted of rearrangements of the elements in different combinations, without any change in the identities of the elements themselves. The theory worked for hundreds of years, and was confirmed experimentally over a wide range of pressures and temperatures and with many combinations of elements. Only in the twentieth century did we learn that one element could be transformed into one another under the conditions of extremely high pressure and temperature existing in a nuclear bomb or inside a star. That observation didn't completely invalidate the original theory of the immutability of the elements, but it showed that it was only an approximation, valid at ordinary temperatures and pressures.

A satirical drawing of an alchemist's laboratory. H. Cock, after a drawing by Peter Brueghel the Elder (16th century).

Self-Check

A psychic conducts seances in which the spirits of the dead speak to the participants. He says he has special psychic powers not possessed by other people, which allow him to "channel" the communications with the spirits. What part of the scientific method is being violated here? [Answer below.]

The scientific method as described here is an idealization, and should not be understood as a set procedure for doing science. Scientists have as many weaknesses and character flaws as any other group, and it is very common for scientists to try to discredit other people's experiments when the results run contrary to their own favored point of view. Successful science also has more to do with luck, intuition, and creativity than most people realize, and the restrictions of the scientific method do not stifle individuality and self-expression any more than the fugue and sonata forms

---

If only he has the special powers, then his results can never be reproduced.

stifled Bach and Haydn. There is a recent tendency among social scientists to go even further and to deny that the scientific method even exists, claiming that science is no more than an arbitrary social system that determines what ideas to accept based on an in-group's criteria. I think that's going too far. If science is an arbitrary social ritual, it would seem difficult to explain its effectiveness in building such useful items as airplanes, CD players and sewers. If alchemy and astrology were no less scientific in their methods than chemistry and astronomy, what was it that kept them from producing anything useful?

Discussion Questions

Consider whether or not the scientific method is being applied in the following examples. If the scientific method is not being applied, are the people whose actions are being described performing a useful human activity, albeit an unscientific one?

A. Acupuncture is a traditional medical technique of Asian origin in which small needles are inserted in the patient's body to relieve pain. Many doctors trained in the west consider acupuncture unworthy of experimental study because if it had therapeutic effects, such effects could not be explained by their theories of the nervous system. Who is being more scientific, the western or eastern practitioners?

B. Goethe, a famous German poet, is less well known for his theory of color. He published a book on the subject, in which he argued that scientific apparatus for measuring and quantifying color, such as prisms, lenses and colored filters, could not give us full insight into the ultimate meaning of color, for instance the cold feeling evoked by blue and green or the heroic sentiments inspired by red. Was his work scientific?

C. A child asks why things fall down, and an adult answers "because of gravity." The ancient Greek philosopher Aristotle explained that rocks fell because it was their nature to seek out their natural place, in contact with the earth. Are these explanations scientific?

D. Buddhism is partly a psychological explanation of human suffering, and psychology is of course a science. The Buddha could be said to have engaged in a cycle of theory and experiment, since he worked by trial and error, and even late in his life he asked his followers to challenge his ideas. Buddhism could also be considered reproducible, since the Buddha told his followers they could find enlightenment for themselves if they followed a certain course of study and discipline. Is Buddhism a scientific pursuit?

0.2 What Is Physics?

Given for one instant an intelligence which could comprehend all the forces by which nature is animated and the respective positions of the things which compose it...nothing would be uncertain, and the future as the past would be laid out before its eyes.

Pierre Simon de Laplace

Physics is the use of the scientific method to find out the basic principles governing light and matter, and to discover the implications of those laws. Part of what distinguishes the modern outlook from the ancient mindset is the assumption that there are rules by which the universe functions, and that those laws can be at least partially understood by humans. From the Age of Reason through the nineteenth century, many scientists began to be convinced that the laws of nature not only could be known but, as claimed by Laplace, those laws could in principle be used to predict everything about the universe's future if complete information was available about the present state of all light and matter. In subsequent sections, I'll describe two general types of limitations on prediction using the laws of physics, which were only recognized in the twentieth century.

Matter can be defined as anything that is affected by gravity, i.e. that has weight or would have weight if it was near the Earth or another star or planet massive enough to produce measurable gravity. Light can be defined as anything that can travel from one place to another through empty space and can influence matter, but has no weight. For example, sunlight can influence your body by heating it or by damaging your DNA and giving you skin cancer. The physicist's definition of light includes a variety of phenomena that are not visible to the eye, including radio waves, microwaves, x-rays, and gamma rays. These are the "colors" of light that do not happen to fall within the narrow violet-to-red range of the rainbow that we can see.

Self-check

At the turn of the 20th century, a strange new phenomenon was discovered in vacuum tubes: mysterious rays of unknown origin and nature. These rays are the same as the ones that shoot from the back of your TV's picture tube and hit the front to make the picture. Physicists in 1895 didn't have the faintest idea what the rays were, so they simply named them "cathode rays," after the name for the electrical contact from which they sprang. A fierce debate raged, complete with nationalistic overtones, over whether the rays were a form of light or of matter. What would they have had to do in order to settle the issue?

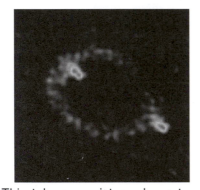

This telescope picture shows two images of the same distant object, an exotic, very luminous object called a quasar. This is interpreted as evidence that a massive, dark object, possibly a black hole, happens to be between us and it. Light rays that would otherwise have missed the earth on either side have been bent by the dark object's gravity so that they reach us. The actual direction to the quasar is presumably in the center of the image, but the light along that central line doesn't get to us because it is absorbed by the dark object. The quasar is known by its catalog number, MG1131+0456, or more informally as Einstein's Ring.

Many physical phenomena are not themselves light or matter, but are properties of light or matter or interactions between light and matter. For instance, motion is a property of all light and some matter, but it is not itself light or matter. The pressure that keeps a bicycle tire blown up is an interaction between the air and the tire. Pressure is not a form of matter in and of itself. It is as much a property of the tire as of the air. Analogously, sisterhood and employment are relationships among people but are not people themselves.

Some things that appear weightless actually do have weight, and so qualify as matter. Air has weight, and is thus a form of matter even though a cubic inch of air weighs less than a grain of sand. A helium balloon has weight, but is kept from falling by the force of the surrounding more dense air, which pushes up on it. Astronauts in orbit around the Earth have weight, and are falling along a curved arc, but they are moving so fast that the curved arc of their fall is broad enough to carry them all the way around the Earth in a circle. They perceive themselves as being weightless because their space capsule is falling along with them, and the floor therefore does not push up on their feet.

Optional Topic
Einstein predicted as a consequence of his theory of relativity that light would after all be affected by gravity, although the effect would be extremely weak under normal conditions. His prediction was borne out by observations of the bending of light rays from stars as they passed close to the sun on their way to the Earth. Einstein also

---

 They would have had to weigh the rays, or check for a loss of weight in the object from which they were have emitted. (For technical reasons, this was not a measurement they could actually do, hence the opportunity for disagreement.)

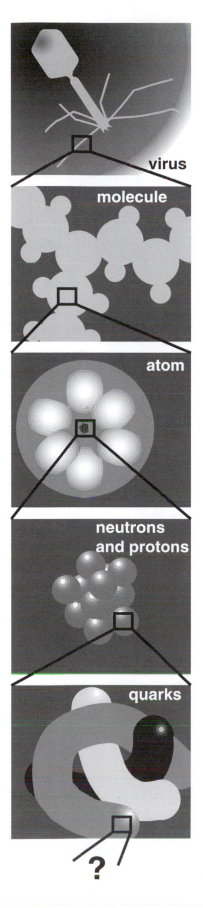

virus

molecule

atom

neutrons and protons

quarks

?

predicted the existence of black holes, stars so massive and compact that their intense gravity would not even allow light to escape. (These days there is strong evidence that black holes exist.)

Einstein's interpretation was that light doesn't really have mass, but that energy is affected by gravity just like mass is. The energy in a light beam is equivalent to a certain amount of mass, given by the famous equation $E=mc^2$, where c is the speed of light. Because the speed of light is such a big number, a large amount of energy is equivalent to only a very small amount of mass, so the gravitational force on a light ray can be ignored for most practical purposes.

There is however a more satisfactory and fundamental distinction between light and matter, which should be understandable to you if you have had a chemistry course. In chemistry, one learns that electrons obey the Pauli exclusion principle, which forbids more than one electron from occupying the same orbital if they have the same spin. The Pauli exclusion principle is obeyed by the subatomic particles of which matter is composed, but disobeyed by the particles, called photons, of which a beam of light is made.

Einstein's theory of relativity is discussed more fully in book 6 of this series.

The boundary between physics and the other sciences is not always clear. For instance, chemists study atoms and molecules, which are what matter is built from, and there are some scientists who would be equally willing to call themselves physical chemists or chemical physicists. It might seem that the distinction between physics and biology would be clearer, since physics seems to deal with inanimate objects. In fact, almost all physicists would agree that the basic laws of physics that apply to molecules in a test tube work equally well for the combination of molecules that constitutes a bacterium. (Some might believe that something more happens in the minds of humans, or even those of cats and dogs.) What differentiates physics from biology is that many of the scientific theories that describe living things, while ultimately resulting from the fundamental laws of physics, cannot be rigorously derived from physical principles.

Isolated systems and reductionism

To avoid having to study everything at once, scientists isolate the things they are trying to study. For instance, a physicist who wants to study the motion of a rotating gyroscope would probably prefer that it be isolated from vibrations and air currents. Even in biology, where field work is indispensable for understanding how living things relate to their entire environment, it is interesting to note the vital historical role played by Darwin's study of the Galápagos Islands, which were conveniently isolated from the rest of the world. Any part of the universe that is considered apart from the rest can be called a "system."

Physics has had some of its greatest successes by carrying this process of isolation to extremes, subdividing the universe into smaller and smaller parts. Matter can be divided into atoms, and the behavior of individual atoms can be studied. Atoms can be split apart into their constituent neutrons, protons and electrons. Protons and neutrons appear to be made out of even smaller particles called quarks, and there have even been some claims of experimental evidence that quarks have smaller parts inside them.

This method of splitting things into smaller and smaller parts and studying how those parts influence each other is called reductionism. The hope is that the seemingly complex rules governing the larger units can be better understood in terms of simpler rules governing the smaller units. To appreciate what reductionism has done for science, it is only necessary to examine a 19th-century chemistry textbook. At that time, the existence of atoms was still doubted by some, electrons were not even suspected to exist, and almost nothing was understood of what basic rules governed the way atoms interacted with each other in chemical reactions. Students had to memorize long lists of chemicals and their reactions, and there was no way to understand any of it systematically. Today, the student only needs to remember a small set of rules about how atoms interact, for instance that atoms of one element cannot be converted into another via chemical reactions, or that atoms from the right side of the periodic table tend to form strong bonds with atoms from the left side.

Discussion Questions

A. I've suggested replacing the ordinary dictionary definition of light with a more technical, more precise one that involves weightlessness. It's still possible, though, that the stuff a lightbulb makes, ordinarily called "light," does have some small amount of weight. Suggest an experiment to attempt to measure whether it does.

B. Heat is weightless (i.e. an object becomes no heavier when heated), and can travel across an empty room from the fireplace to your skin, where it influences you by heating you. Should heat therefore be considered a form of light by our definition? Why or why not?

C. Similarly, should sound be considered a form of light?

0.3 How to Learn Physics

For as knowledges are now delivered, there is a kind of contract of error between the deliverer and the receiver; for he that delivereth knowledge desireth to deliver it in such a form as may be best believed, and not as may be best examined; and he that receiveth knowledge desireth rather present satisfaction than expectant inquiry.

Sir Francis Bacon

Many students approach a science course with the idea that they can succeed by memorizing the formulas, so that when a problem is assigned on the homework or an exam, they will be able to plug numbers in to the formula and get a numerical result on their calculator. Wrong! That's not what learning science is about! There is a big difference between memorizing formulas and understanding concepts. To start with, different formulas may apply in different situations. One equation might represent a definition, which is always true. Another might be a very specific equation for the speed of an object sliding down an inclined plane, which would not be true if the object was a rock drifting down to the bottom of the ocean. If you don't work to understand physics on a conceptual level, you won't know which formulas can be used when.

Other Books

PSSC Physics, Haber-Schaim et al., 7th ed., 1986. Kendall/Hunt, Dubuque, Iowa.

A high-school textbook at the algebra-based level. This book distinguishes itself by giving a clear, careful, and honest explanation of every topic, while avoiding unnecessary details.

Physics for Poets, Robert H. March, 4th ed., 1996. McGraw-Hill, New York.

As the name implies, this book's intended audience is liberal arts students who want to understand science in a broader cultural and historical context. Not much math is used, and the page count of this little paperback is about five times less than that of the typical "kitchen sink" textbook, but the intellectual level is actually pretty challenging.

Conceptual Physics, Paul Hewitt. Scott Foresman, Glenview, Ill.

This is the excellent book used for Physics 130 here at Fullerton College. Only simple algebra is used.

Most students taking college science courses for the first time also have very little experience with interpreting the meaning of an equation. Consider the equation $w = A/h$ relating the width of a rectangle to its height and area. A student who has not developed skill at interpretation might view this as yet another equation to memorize and plug in to when needed. A slightly more savvy student might realize that it is simply the familiar formula $A = wh$ in a different form. When asked whether a rectangle would have a greater or smaller width than another with the same area but a smaller height, the unsophisticated student might be at a loss, not having any numbers to plug in on a calculator. The more experienced student would know how to reason about an equation involving division — if h is smaller, and A stays the same, then w must be bigger. Often, students fail to recognize a sequence of equations as a derivation leading to a final result, so they think all the intermediate steps are equally important formulas that they should memorize.

When learning any subject at all, it is important to become as actively involved as possible, rather than trying to read through all the information quickly without thinking about it. It is a good idea to read and think about the questions posed at the end of each section of these notes as you encounter them, so that you know you have understood what you were reading.

Many students' difficulties in physics boil down mainly to difficulties with math. Suppose you feel confident that you have enough mathematical preparation to succeed in this course, but you are having trouble with a few specific things. In some areas, the brief review given in this chapter may be sufficient, but in other areas it probably will not. Once you identify the areas of math in which you are having problems, get help in those areas. Don't limp along through the whole course with a vague feeling of dread about something like scientific notation. The problem will not go away if you ignore it. The same applies to essential mathematical skills that you are learning in this course for the first time, such as vector addition.

Sometimes students tell me they keep trying to understand a certain topic in the book, and it just doesn't make sense. The worst thing you can possibly do in that situation is to keep on staring at the same page. Every textbook explains certain things badly — even mine! — so the best thing to do in this situation is to look at a different book. Instead of college textbooks aimed at the same mathematical level as the course you're taking, you may in some cases find that high school books or books at a lower math level give clearer explanations. The three books listed on the left are, in my opinion, the best introductory physics books available, although they would not be appropriate as the primary textbook for a college-level course for science majors.

Finally, when reviewing for an exam, don't simply read back over the text and your lecture notes. Instead, try to use an active method of reviewing, for instance by discussing some of the discussion questions with another student, or doing homework problems you hadn't done the first time.

0.4   Self-Evaluation

The introductory part of a book like this is hard to write, because every student arrives at this starting point with a different preparation. One student may have grown up in another country and so may be completely comfortable with the metric system, but may have had an algebra course in which the instructor passed too quickly over scientific notation. Another student may have already taken calculus, but may have never learned the metric system. The following self-evaluation is a checklist to help you figure out what you need to study to be prepared for the rest of the course.

If you disagree with this statement...	you should study this section:
I am familiar with the basic metric units of meters, kilograms, and seconds, and the most common metric prefixes: milli- (m), kilo- (k), and centi- (c).	0.5 Basics of the Metric System
I know about the Newton, a unit of force	0.6 The Newton, the Metric Unit of Force
I am familiar with these less common metric prefixes: mega- (M), micro- (μ), and nano- (n).	0.7 Less Common Metric Prefixes
I am comfortable with scientific notation.	0.8 Scientific Notation
I can confidently do metric conversions.	0.9 Conversions
I understand the purpose and use of significant figures.	0.10 Significant Figures

It wouldn't hurt you to skim the sections you think you already know about, and to do the self-checks in those sections.

0.5   Basics of the Metric System

The metric system

Units were not standardized until fairly recently in history, so when the physicist Isaac Newton gave the result of an experiment with a pendulum, he had to specify not just that the string was 37 $^7/_8$ inches long but that it was "37 $^7/_8$ London inches long." The inch as defined in Yorkshire would have been different. Even after the British Empire standardized its units, it was still very inconvenient to do calculations involving money, volume, distance, time, or weight, because of all the odd conversion factors, like 16 ounces in a pound, and 5280 feet in a mile. Through the nineteenth century, schoolchildren squandered most of their mathematical education in preparing to do calculations such as making change when a customer in a shop offered a one-crown note for a book costing two pounds, thirteen shillings and tuppence. The dollar has always been decimal, and British money went decimal decades ago, but the United States is still saddled with the antiquated system of feet, inches, pounds, ounces and so on.

Every country in the world besides the U.S. has adopted a system of units known in English as the "metric system." This system is entirely decimal, thanks to the same eminently logical people who brought about the French Revolution. In deference to France, the system's official name is the Système International, or SI, meaning International System. (The phrase "SI system" is therefore redundant.)

The wonderful thing about the SI is that people who live in countries more modern than ours do not need to memorize how many ounces there are in a pound, how many cups in a pint, how many feet in a mile, etc. The whole system works with a single, consistent set of prefixes (derived from Greek) that modify the basic units. Each prefix stands for a power of ten, and has an abbreviation that can be combined with the symbol for the unit. For instance, the meter is a unit of distance. The prefix kilo- stands for 10^3, so a kilometer, 1 km, is a thousand meters.

The basic units of the metric system are the meter for distance, the second for time, and the gram for mass.

The following are the most common metric prefixes. You should memorize them.

prefix		meaning	example	
kilo-	k	10^3	60 kg	= a person's mass
centi-	c	10^{-2}	28 cm	= height of a piece of paper
milli-	m	10^{-3}	1 ms	= time for one vibration of a guitar string playing the note D

The prefix centi-, meaning 10^{-2}, is only used in the centimeter; a hundredth of a gram would not be written as 1 cg but as 10 mg. The centi-prefix can be easily remembered because a cent is 10^{-2} dollars. The official SI abbreviation for seconds is "s" (not "sec") and grams are "g" (not "gm").

The second

> The sun stood still and the moon halted until the nation had taken vengeance on its enemies...
>
> Joshua 10:12-14
>
> Absolute, true, and mathematical time, of itself, and from its own nature, flows equably without relation to anything external...
>
> Isaac Newton

When I stated briefly above that the second was a unit of time, it may not have occurred to you that this was not really much of a definition. The two quotes above are meant to demonstrate how much room for confusion exists among people who seem to mean the same thing by a word such as "time." The first quote has been interpreted by some biblical scholars as indicating an ancient belief that the motion of the sun across the sky was not just something that occurred with the passage of time but that the sun actually caused time to pass by its motion, so that freezing it in the sky

The Time Without Underwear

Unfortunately, the French Revolutionary calendar never caught on. Each of its twelve months was 30 days long, with names like Thermidor (the month of heat) and Germinal (the month of budding). To round out the year to 365 days, a five-day period was added on the end of the calendar, and named the *sans culottides*. In modern French, *sans culottides* means "time without underwear," but in the 18th century, it was a way to honor the workers and peasants, who wore simple clothing instead of the fancy pants (*culottes*) of the aristocracy.

Pope Gregory created our modern "Gregorian" calendar, with its system of leap years, to make the length of the calendar year match the length of the cycle of seasons. Not until 1752 did Protestant England switched to the new calendar. Some less educated citizens believed that the shortening of the month by eleven days would shorten their lives by the same interval. In this illustration by William Hogarth, the leaflet lying on the ground reads, "Give us our eleven days."

would have some kind of a supernatural decelerating effect on everyone except the Hebrew soldiers. Many ancient cultures also conceived of time as cyclical, rather than proceeding along a straight line as in 1998, 1999, 2000, 2001,... The second quote, from a relatively modern physicist, may sound a lot more scientific, but most physicists today would consider it useless as a definition of time. Today, the physical sciences are based on *operational definitions*, which means definitions that spell out the actual steps (operations) required to measure something numerically.

Now in an era when our toasters, pens, and coffee pots tell us the time, it is far from obvious to most people what is the fundamental operational definition of time. Until recently, the hour, minute, and second were defined operationally in terms of the time required for the earth to rotate about its axis. Unfortunately, the Earth's rotation is slowing down slightly, and by 1967 this was becoming an issue in scientific experiments requiring precise time measurements. The second was therefore redefined as the time required for a certain number of vibrations of the light waves emitted by a cesium atoms in a lamp constructed like a familiar neon sign but with the neon replaced by cesium. The new definition not only promises to stay constant indefinitely, but for scientists is a more convenient way of calibrating a clock than having to carry out astronomical measurements.

Self-Check

What is a possible operational definition of how strong a person is?

The meter

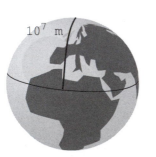

10^7 m

The French originally defined the meter as 10^{-7} times the distance from the equator to the north pole, as measured through Paris (of course). Even if the definition was operational, the operation of traveling to the north pole and laying a surveying chain behind you was not one that most working scientists wanted to carry out. Fairly soon, a standard was created in the form of a metal bar with two scratches on it. This definition persisted until 1960, when the meter was redefined as the distance traveled by light in a vacuum over a period of (1/299792458) seconds.

A dictionary might define "strong" as "posessing powerful muscles," but that's not an operational definition, because it doesn't say how to measure strength numerically. One possible operational definition would be the number of pounds a person can bench press.

The kilogram

The third base unit of the SI is the kilogram, a unit of mass. Mass is intended to be a measure of the amount of a substance, but that is not an operational definition. Bathroom scales work by measuring our planet's gravitational attraction for the object being weighed, but using that type of scale to define mass operationally would be undesirable because gravity varies in strength from place to place on the earth.

There's a surprising amount of disagreement among physics textbooks about how mass should be defined, but here's how it's actually handled by the few working physicists who specialize in ultra-high-precision measurements. They maintain a physical object in Paris, which is the standard kilogram, a cylinder made of platinum-iridium alloy. Duplicates are checked against this mother of all kilograms by putting the original and the copy on the two opposite pans of a balance. Although this method of comparison depends on gravity, the problems associated with differences in gravity in different geographical locations are bypassed, because the two objects are being compared in the same place. The duplicates can then be removed from the Parisian kilogram shrine and transported elsewhere in the world.

Combinations of metric units

Just about anything you want to measure can be measured with some combination of meters, kilograms, and seconds. Speed can be measured in m/s, volume in m³, and density in kg/m³. Part of what makes the SI great is this basic simplicity. No more funny units like a cord of wood, a bolt of cloth, or a jigger of whiskey. No more liquid and dry measure. Just a simple, consistent set of units. The SI measures put together from meters, kilograms, and seconds make up the mks system. For example, the mks unit of speed is m/s, not km/hr.

Discussion question

Isaac Newton wrote, "...the natural days are truly unequal, though they are commonly considered as equal, and used for a measure of time... It may be that there is no such thing as an equable motion, whereby time may be accurately measured. All motions may be accelerated or retarded..." Newton was right. Even the modern definition of the second in terms of light emitted by cesium atoms is subject to variation. For instance, magnetic fields could cause the cesium atoms to emit light with a slightly different rate of vibration. What makes us think, though, that a pendulum clock is more accurate than a sundial, or that a cesium atom is a more accurate timekeeper than a pendulum clock? That is, how can one test experimentally how the accuracies of different time standards compare?

0.6   The Newton, the Metric Unit of Force

A force is a push or a pull, or more generally anything that can change an object's speed or direction of motion. A force is required to start a car moving, to slow down a baseball player sliding in to home base, or to make an airplane turn. (Forces may fail to change an object's motion if they are canceled by other forces, e.g. the force of gravity pulling you down right now is being canceled by the force of the chair pushing up on you.) The metric unit of force is the Newton, defined as the force which, if applied for one second, will cause a 1-kilogram object starting from rest to reach a speed of 1 m/s. Later chapters will discuss the force concept in more detail. In fact, this entire book is about the relationship between force and motion.

In the previous section, I gave a gravitational definition of mass, but by defining a numerical scale of force, we can also turn around and define a scale of mass without reference to gravity. For instance, if a force of two Newtons is required to accelerate a certain object from rest to 1 m/s in 1 s, then that object must have a mass of 2 kg. From this point of view, mass characterizes an object's resistance to a change in its motion, which we call inertia or inertial mass. Although there is no fundamental reason why an object's resistance to a change in its motion must be related to how strongly gravity affects it, careful and precise experiments have shown that the inertial definition and the gravitational definition of mass are highly consistent for a variety of objects. It therefore doesn't really matter for any practical purpose which definition one adopts.

Discussion Question

Spending a long time in weightlessness is unhealthy. One of the most important negative effects experienced by astronauts is a loss of muscle and bone mass. Since an ordinary scale won't work for an astronaut in orbit, what is a possible way of monitoring this change in mass? (Measuring the astronaut's waist or biceps with a measuring tape is not good enough, because it doesn't tell anything about bone mass, or about the replacement of muscle with fat.)

0.7 Less Common Metric Prefixes

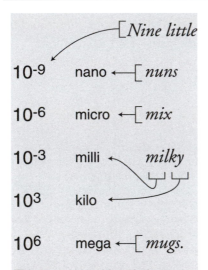

This is a mnemonic to help you remember the most important metric prefixes. The word "little" is to remind you that the list starts with the prefixes used for small quantities and builds upward. The exponent changes by 3 with each step, except that of course we do not need a special prefix for 10^0, which equals one.

The following are three metric prefixes which, while less common than the ones discussed previously, are well worth memorizing.

prefix		meaning	example
mega-	M	10^6	6.4 Mm = radius of the earth
micro-	μ	10^{-6}	1 μm = diameter of a human hair
nano-	n	10^{-9}	0.154 nm = distance between carbon nuclei in an ethane molecule

Note that the abbreviation for micro is the Greek letter mu, μ — a common mistake is to confuse it with m (milli) or M (mega).

There are other prefixes even less common, used for extremely large and small quantities. For instance, 1 femtometer=10^{-15} m is a convenient unit of distance in nuclear physics, and 1 gigabyte=10^9 bytes is used for computers' hard disks. The international committee that makes decisions about the SI has recently even added some new prefixes that sound like jokes, e.g. 1 yoctogram = 10^{-24} g is about half the mass of a proton. In the immediate future, however, you're unlikely to see prefixes like "yocto-" and "zepto-" used except perhaps in trivia contests at science-fiction conventions or other geekfests.

Suppose you could slow down time so that according to your perception, a beam of light would move across a room at the speed of a slow walk. If you perceived a nanosecond as if it was a second, how would you perceive a microsecond?

0.8 Scientific Notation

Most of the interesting phenomena our universe has to offer are not on the human scale. It would take about 1,000,000,000,000,000,000,000 bacteria to equal the mass of a human body. When the physicist Thomas Young discovered that light was a wave, it was back in the bad old days before scientific notation, and he was obliged to write that the time required for one vibration of the wave was 1/500 of a millionth of a millionth of a second. Scientific notation is a less awkward way to write very large and very small numbers such as these. Here's a quick review.

Scientific notation means writing a number in terms of a product of something from 1 to 10 and something else that is a power of ten. For instance,

$$32 = 3.2 \times 10^1$$
$$320 = 3.2 \times 10^2$$
$$3200 = 3.2 \times 10^3 \dots$$

Each number is ten times bigger than the previous one.

Since 10^1 is ten times smaller than 10^2, it makes sense to use the notation 10^0 to stand for one, the number that is in turn ten times smaller than 10^1. Continuing on, we can write 10^{-1} to stand for 0.1, the number ten times smaller than 10^0. Negative exponents are used for small numbers:

$$3.2 = 3.2 \times 10^0$$
$$0.32 = 3.2 \times 10^{-1}$$
$$0.032 = 3.2 \times 10^{-2} \dots$$

A common source of confusion is the notation used on the displays of many calculators. Examples:

3.2×10^6 (written notation)

$3.2E+6$ (notation on some calculators)

3.2^6 (notation on some other calculators)

The last example is particularly unfortunate, because 3.2^6 really stands for the number $3.2 \times 3.2 \times 3.2 \times 3.2 \times 3.2 \times 3.2 = 1074$, a totally different number from $3.2 \times 10^6 = 3200000$. The calculator notation should never be used in writing. It's just a way for the manufacturer to save money by making a simpler display.

  A microsecond is 1000 times longer than a nanosecond, so it would seem like 1000 seconds, or about 20 minutes.

A student learns that 10^4 bacteria, standing in line to register for classes at Paramecium Community College, would form a queue of this size:

_____

The student concludes that 10^2 bacteria would form a line of this length:

_____

Why is the student incorrect?

0.9 Conversions

I suggest you avoid memorizing lots of conversion factors between SI units and U.S. units. Suppose the United Nations sends its black helicopters to invade California (after all who wouldn't rather live here than in New York City?), and institutes water fluoridation and the SI, making the use of inches and pounds into a crime punishable by death. I think you could get by with only two mental conversion factors:

1 inch = 2.54 cm
An object with a weight on Earth of 2.2 lb has a mass of 1 kg.

The first one is the present definition of the inch, so it's exact. The second one is not exact, but is good enough for most purposes. The pound is a unit of gravitational force, while the kg is a unit of mass, which measures how hard it is to accelerate an object, not how hard gravity pulls on it. Therefore it would be incorrect to say that 2.2 lb literally equaled 1 kg, even approximately.

More important than memorizing conversion factors is understanding the right method for doing conversions. Even within the SI, you may need to convert, say, from grams to kilograms. Different people have different ways of thinking about conversions, but the method I'll describe here is systematic and easy to understand. The idea is that if 1 kg and 1000 g represent the same mass, then we can consider a fraction like

$$\frac{10^3 \, \text{g}}{1 \, \text{kg}}$$

to be a way of expressing the number one. This may bother you. For instance, if you type 1000/1 into your calculator, you will get 1000, not one. Again, different people have different ways of thinking about it, but the justification is that it helps us to do conversions, and it works! Now if we want to convert 0.7 kg to units of grams, we can multiply 0.7 kg by the number one:

$$0.7 \, \text{kg} \times \frac{10^3 \, \text{g}}{1 \, \text{kg}}$$

If you're willing to treat symbols such as "kg" as if they were variables as used in algebra (which they're really not), you can then cancel the kg on top with the kg on the bottom, resulting in

---

Exponents have to do with multiplication, not addition. The first line should be 100 times longer than the second, not just twice as long.

$$0.7 \, \cancel{kg} \times \frac{10^3 \, g}{1 \, \cancel{kg}} = 700 \, g \quad .$$

To convert grams to kilograms, you would simply flip the fraction upside down.

One advantage of this method is that it can easily be applied to a series of conversions. For instance, to convert one year to units of seconds,

$$1 \, \cancel{year} \times \frac{365 \, \cancel{days}}{1 \, \cancel{year}} \times \frac{24 \, \cancel{hours}}{1 \, \cancel{day}} \times \frac{60 \, \cancel{min}}{1 \, \cancel{hour}} \times \frac{60 \, s}{1 \, \cancel{min}}$$

$$= 3.15 \times 10^7 \, s \quad .$$

Should that exponent be positive or negative?

A common mistake is to write the conversion fraction incorrectly. For instance the fraction

$$\frac{10^3 \, kg}{1 \, g} \qquad \text{(incorrect)}$$

does not equal one, because 10^3 kg is the mass of a car, and 1 g is the mass of a raisin. One correct way of setting up the conversion factor would be

$$\frac{10^{-3} \, kg}{1 \, g} \quad . \qquad \text{(correct)}$$

You can usually detect such a mistake if you take the time to check your answer and see if it is reasonable.

If common sense doesn't rule out either a positive or a negative exponent, here's another way to make sure you get it right. There are big prefixes and small prefixes:

> big prefixes: k   M
> small prefixes: m   μ   n

(It's not hard to keep straight which are which, since "mega" and "micro" are evocative, and it's easy to remember that a kilometer is bigger than a meter and a millimeter is smaller.) In the example above, we want the top of the fraction to be the same as the bottom. Since k is a big prefix, we need to *compensate* by putting a small number like 10^{-3} in front of it, not a big number like 10^3.

Discussion Question

Each of the following conversions contains an error. In each case, explain what the error is.

(a) $1000 \, kg \times \dfrac{1 \, kg}{1000 \, g} = 1 \, g$ (b) $50 \, m \times \dfrac{1 \, cm}{100 \, m} = 0.5 \, cm$

(c) "Nano" is 10^{-9}, so there are 10^{-9} nm in a meter.

(d) "Micro" is 10^{-6}, so 1 kg is 10^6 μg.

0.10   Significant Figures

An engineer is designing a car engine, and has been told that the diameter of the pistons (which are being designed by someone else) is 5 cm. He knows that 0.02 cm of clearance is required for a piston of this size, so he designs the cylinder to have an inside diameter of 5.04 cm. Luckily, his supervisor catches his mistake before the car goes into production. She explains his error to him, and mentally puts him in the "do not promote" category.

What was his mistake? The person who told him the pistons were 5 cm in diameter was wise to the ways of significant figures, as was his boss, who explained to him that he needed to go back and get a more accurate number for the diameter of the pistons. That person said "5 cm" rather than "5.00 cm" specifically to avoid creating the impression that the number was extremely accurate. In reality, the pistons' diameter was 5.13 cm. They would never have fit in the 5.04-cm cylinders.

The number of digits of accuracy in a number is referred to as the number of significant figures, or "sig figs" for short. As in the example above, sig figs provide a way of showing the accuracy of a number. In most cases, the result of a calculation involving several pieces of data can be no more accurate than the least accurate piece of data. In other words, "garbage in, garbage out." Since the 5 cm diameter of the pistons was not very accurate, the result of the engineer's calculation, 5.04 cm, was really not as accurate as he thought. In general, your result should not have more than the number of sig figs in the least accurate piece of data you started with. The calculation above should have been done as follows:

5 cm	(1 sig fig)
+ 0.04 cm	(1 sig fig)
= 5 cm	(rounded off to 1 sig fig)

The fact that the final result only has one significant figure then alerts you to the fact that the result is not very accurate, and would not be appropriate for use in designing the engine.

Note that the leading zeroes in the number 0.04 do not count as significant figures, because they are only placeholders. On the other hand, a number such as 50 cm is ambiguous — the zero could be intended as a significant figure, or it might just be there as a placeholder. The ambiguity involving trailing zeroes can be avoided by using scientific notation, in which 5×10^1 cm would imply one sig fig of accuracy, while 5.0×10^1 cm would imply two sig figs.

Self-Check

(a) The following quote is taken from an editorial by Norimitsu Onishi in the New York Times, August 18, 2002.

> Consider Nigeria. Everyone agrees it is Africa's most populous nation. But what is its population? The United Nations says 114 million; the State Department, 120 million. The World Bank says 126.9 million, while the Central Intelligence Agency puts it at 126,635,626.

What should bother you about this?

The various estimates differ by 5 to 10 million. The CIA's estimate includes a ridiculous number of gratuitous significant figures. Does the CIA understand that every day, people in are born in, die in, immigrate to, and emigrate from Nigeria?

Dealing correctly with significant figures can save you time! Often, students copy down numbers from their calculators with eight significant figures of precision, then type them back in for a later calculation. That's a waste of time, unless your original data had that kind of incredible precision.

The rules about significant figures are only rules of thumb, and are not a substitute for careful thinking. For instance, $20.00 + $0.05 is $20.05. It need not and should not be rounded off to $20. In general, the sig fig rules work best for multiplication and division, and we also apply them when doing a complicated calculation that involves many types of operations. For simple addition and subtraction, it makes more sense to maintain a fixed number of digits after the decimal point.

When in doubt, don't use the sig fig rules at all. Instead, intentionally change one piece of your initial data by the maximum amount by which you think it could have been off, and recalculate the final result. The digits on the end that arecompletely reshuffled are the ones that are meaningless, and should be omitted.

How many significant figures are there in each of the following measurements?
(a) 9.937 m
(b) 4.0 s
(c) 0.0000037 kg

(a) (b) 4; (c) 2; (d) 2

Summary

Selected Vocabulary

matter	Anything that is affected by gravity.
light..................................	Anything that can travel from one place to another through empty space and can influence matter, but is not affected by gravity.
operational definition	A definition that states what operations should be carried out to measure the thing being defined.
Système International	A fancy name for the metric system.
mks system	The use of metric units based on the meter, kilogram, and second. Example: meters per second is the mks unit of speed, not cm/s or km/hr.
mass	A numerical measure of how difficult it is to change an object's motion.
significant figures	Digits that contribute to the accuracy of a measurement.

Notation

m	symbol for mass, or the meter, the metric distance unit
kg	kilogram, the metric unit of mass
s	second, the metric unit of time
M-	the metric prefix mega-, 10^6
k-	the metric prefix kilo-, 10^3
m-	the metric prefix milli-, 10^{-3}
μ-	the metric prefix micro-, 10^{-6}
n-	the metric prefix nano-, 10^{-9}

Summary

Physics is the use of the scientific method to study the behavior of light and matter. The scientific method requires a cycle of theory and experiment, theories with both predictive and explanatory value, and reproducible experiments.

The metric system is a simple, consistent framework for measurement built out of the meter, the kilogram, and the second plus a set of prefixes denoting powers of ten. The most systematic method for doing conversions is shown in the following example:

$$370 \text{ ms} \times \frac{10^{-3} \text{ s}}{1 \text{ ms}} = 0.37 \text{ s}$$

Mass is a measure of the amount of a substance. Mass can be defined gravitationally, by comparing an object to a standard mass on a double-pan balance, or in terms of inertia, by comparing the effect of a force on an object to the effect of the same force on a standard mass. The two definitions are found experimentally to be proportional to each other to a high degree of precision, so we usually refer simply to "mass," without bothering to specify which type.

A force is that which can change the motion of an object. The metric unit of force is the Newton, defined as the force required to accelerate a standard 1-kg mass from rest to a speed of 1 m/s in 1 s.

Scientific notation means, for example, writing 3.2×10^5 rather than 320000.

Writing numbers with the correct number of significant figures correctly communicates how accurate they are. As a rule of thumb, the final result of a calculation is no more accurate than, and should have no more significant figures than, the least accurate piece of data.

Homework Problems

1. Correct use of a calculator: (a✔) Calculate $\dfrac{74658}{53222 + 97554}$ on a calculator.

[Self-check: The most common mistake results in 97555.40.]

(b) Which would be more like the price of a TV, and which would be more like the price of a house, $ 3.5x10^5$ or $ 3.5^5$?

2. Compute the following things. If they don't make sense because of units, say so.

(a) 3 cm + 5 cm

(b) 1.11 m + 22 cm

(c) 120 miles + 2.0 hours

(d) 120 miles / 2.0 hours

3. Your backyard has brick walls on both ends. You measure a distance of 23.4 m from the inside of one wall to the inside of the other. Each wall is 29.4 cm thick. How far is it from the outside of one wall to the outside of the other? Pay attention to significant figures.

4 ✓. The speed of light is $3.0x10^8$ m/s. Convert this to furlongs per fortnight. A furlong is 220 yards, and a fortnight is 14 days. An inch is 2.54 cm.

5 ✓. Express each of the following quantities in micrograms: (a) 10 mg, (b) 10^4 g, (c) 10 kg, (d) $100x10^3$ g, (e) 1000 ng.

6 S. Convert 134 mg to units of kg, writing your answer in scientific notation.

7✓. In the last century, the average age of the onset of puberty for girls has decreased by several years. Urban folklore has it that this is because of hormones fed to beef cattle, but it is more likely to be because modern girls have more body fat on the average and possibly because of estrogen-mimicking chemicals in the environment from the breakdown of pesticides. A hamburger from a hormone-implanted steer has about 0.2 ng of estrogen (about double the amount of natural beef). A serving of peas contains about 300 ng of estrogen. An adult woman produces about 0.5 mg of estrogen per day (note the different unit!). (a) How many hamburgers would a girl have to eat in one day to consume as much estrogen as an adult woman's daily production? (b) How many servings of peas?

8 S. The usual definition of the mean (average) of two numbers a and b is $(a+b)/2$. This is called the arithmetic mean. The geometric mean, however, is defined as $(ab)^{1/2}$. For the sake of definiteness, let's say both numbers have units of mass. (a) Compute the arithmetic mean of two numbers that have units of grams. Then convert the numbers to units of kilograms and recompute their mean. Is the answer consistent? (b) Do the same for the geometric mean. (c) If a and b both have units of grams, what should we call the units of ab? Does your answer make sense when you take the square root? (d) Suppose someone proposes to you a third kind of mean, called the superduper mean, defined as $(ab)^{1/3}$. Is this reasonable?

S   A solution is given in the back of the book.   ★ A difficult problem.

✓ A computerized answer check is available.   ∫   A problem that requires calculus.

Life would be very different if you were the size of an insect.

1 Scaling and Order-of-Magnitude Estimates

1.1 Introduction

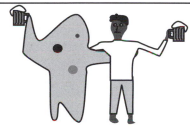

Amoebas this size are seldom encountered.

Why can't an insect be the size of a dog? Some skinny stretched-out cells in your spinal cord are a meter tall — why does nature display no single cells that are not just a meter tall, but a meter wide, and a meter thick as well? Believe it or not, these are questions that can be answered fairly easily without knowing much more about physics than you already do. The only mathematical technique you really need is the humble conversion, applied to area and volume.

Area and volume

Area can be defined by saying that we can copy the shape of interest onto graph paper with 1 cm x 1 cm squares and count the number of squares inside. Fractions of squares can be estimated by eye. We then say the area equals the number of squares, in units of square cm. Although this might seem less "pure" than computing areas using formulae like $A=\pi r^2$ for a circle or $A=wh/2$ for a triangle, those formulae are not useful as definitions of area because they cannot be applied to irregularly shaped areas.

Units of square cm are more commonly written as cm^2 in science. Of course, the unit of measurement symbolized by "cm" is not an algebra symbol standing for a number that can be literally multiplied by itself. But it is advantageous to write the units of area that way and treat the units as if they were algebra symbols. For instance, if you have a rectangle with an area of 6 m^2 and a width of 2 m, then calculating its length as $(6\ m^2)/(2\ m)=3$ m gives a result that makes sense both numerically and in terms of units. This algebra-style treatment of the units also ensures that our methods of converting units work out correctly. For instance, if we accept the fraction

$$\frac{100 \text{ cm}}{1 \text{ m}}$$

as a valid way of writing the number one, then one times one equals one, so we should also say that one can be represented by

$$\frac{100 \text{ cm}}{1 \text{ m}} \times \frac{100 \text{ cm}}{1 \text{ m}}$$

which is the same as

$$\frac{10000 \text{ cm}^2}{1 \text{ m}^2} .$$

That means the conversion factor from square meters to square centimeters is a factor of 10^4, i.e. a square meter has 10^4 square centimeters in it.

All of the above can be easily applied to volume as well, using one-cubic-centimeter blocks instead of squares on graph paper.

To many people, it seems hard to believe that a square meter equals 10000 square centimeters, or that a cubic meter equals a million cubic centimeters — they think it would make more sense if there were 100 cm^2 in 1 m^2, and 100 cm^3 in 1 m^3, but that would be incorrect. The examples shown in the figure below aim to make the correct answer more believable, using the traditional U.S. units of feet and yards. (One foot is 12 inches, and one yard is three feet.)

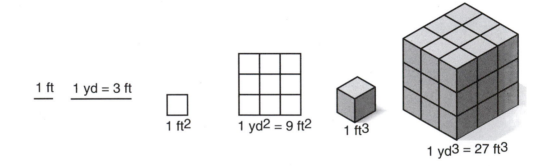

1 ft   1 yd = 3 ft

1 ft²   1 yd² = 9 ft²   1 ft³   1 yd³ = 27 ft³

Self-Check

Based on the figure, convince yourself that there are 9 ft² in a square yard , and 27 ft³ in a cubic yard, then demonstrate the same thing symbolically (i.e. with the method using fractions that equal one).

Discussion question

A. How many square centimeters are there in a square inch? (1 inch=2.54 cm) First find an approximate answer by making a drawing, then derive the conversion factor more accurately using the symbolic method.

1 yd²x(3 ft/1 yd)²=9 ft².   1 yd³x(3 ft/1 yd)³=27 ft³.

Galileo Galilei (1564-1642) was a Renaissance Italian who brought the scientific method to bear on physics, creating the modern version of the science. Coming from a noble but very poor family, Galileo had to drop out of medical school at the University of Pisa when he ran out of money. Eventually becoming a lecturer in mathematics at the same school, he began a career as a notorious troublemaker by writing a burlesque ridiculing the university's regulations — he was forced to resign, but found a new teaching position at Padua. He invented the pendulum clock, investigated the motion of falling bodies, and discovered the moons of Jupiter. The thrust of his life's work was to discredit Aristotle's physics by confronting it with contradictory experiments, a program which paved the way for Newton's discovery of the relationship between force and motion. In Chapter 3 we'll come to the story of Galileo's ultimate fate at the hands of the Church.

1.2   Scaling of Area and Volume

Great fleas have lesser fleas
Upon their backs to bite 'em.
And lesser fleas have lesser still,
And so ad infinitum.

Jonathan Swift

The small boat holds up just fine.

A larger boat built with the same proportions as the small one will collapse under its own weight.

A boat this large needs to have timbers that are thicker compared to its size.

Now how do these conversions of area and volume relate to the questions I posed about sizes of living things? Well, imagine that you are shrunk like Alice in Wonderland to the size of an insect. One way of thinking about the change of scale is that what used to look like a centimeter now looks like perhaps a meter to you, because you're so much smaller. If area and volume scaled according to most people's intuitive, incorrect expectations, with 1 m^2 being the same as 100 cm^2, then there would be no particular reason why nature should behave any differently on your new, reduced scale. But nature does behave differently now that you're small. For instance, you will find that you can walk on water, and jump to many times your own height. The physicist Galileo Galilei had the basic insight that the scaling of area and volume determines how natural phenomena behave differently on different scales. He first reasoned about mechanical structures, but later extended his insights to living things, taking the then-radical point of view that at the fundamental level, a living organism should follow the same laws of nature as a machine. We will follow his lead by first discussing machines and then living things.

Galileo on the behavior of nature on large and small scales

One of the world's most famous pieces of scientific writing is Galileo's **Dialogues Concerning the Two New Sciences**. Galileo was an entertaining writer who wanted to explain things clearly to laypeople, and he livened up his work by casting it in the form of a dialogue among three people. Salviati is really Galileo's alter ego. Simplicio is the stupid character, and one of the reasons Galileo got in trouble with the Church was that there were rumors that Simplicio represented the Pope. Sagredo is the earnest and intelligent student, with whom the reader is supposed to identify. (The following excerpts are from the 1914 translation by Crew and de Salvio.)

This plank is the longest it can be without collapsing under its own weight. If it was a hundredth of an inch longer, it would collapse.

This plank is made out of the same kind of wood. It is twice as thick, twice as long, and twice as wide. It will collapse under its own weight.

(After Galileo's original drawing.)

SALVIATI: ...we asked the reason why [shipbuilders] employed stocks, scaffolding, and bracing of larger dimensions for launching a big vessel than they do for a small one; and [an old man] answered that they did this in order to avoid the danger of the ship parting under its own heavy weight, a danger to which small boats are not subject?

SAGREDO: Yes, that is what I mean; and I refer especially to his last assertion which I have always regarded as false...; namely, that in speaking of these and other similar machines one cannot argue from the small to the large, because many devices which succeed on a small scale do not work on a large scale. Now, since mechanics has its foundations in geometry, where mere size [is unimportant], I do not see that the properties of circles, triangles, cylinders, cones and other solid figures will change with their size. If, therefore, a large machine be constructed in such a way that its parts bear to one another the same ratio as in a smaller one, and if the smaller is sufficiently strong for the purpose for which it is designed, I do not see why the larger should not be able to withstand any severe and destructive tests to which it may be subjected.

Salviati contradicts Sagredo:

SALVIATI: ...Please observe, gentlemen, how facts which at first seem improbable will, even on scant explanation, drop the cloak which has hidden them and stand forth in naked and simple beauty. Who does not know that a horse falling from a height of three or four cubits will break his bones, while a dog falling from the same height or a cat from a height of eight or ten cubits will suffer no injury? Equally harmless would be the fall of a grasshopper from a tower or the fall of an ant from the distance of the moon.

The point Galileo is making here is that small things are sturdier in proportion to their size. There are a lot of objections that could be raised, however. After all, what does it really mean for something to be "strong", to be "strong in proportion to its size," or to be strong "out of proportion to its size?" Galileo hasn't spelled out operational definitions of things like "strength," i.e. definitions that spell out how to measure them numerically.

Also, a cat is shaped differently from a horse — an enlarged photograph of a cat would not be mistaken for a horse, even if the photo-doctoring experts at the National Inquirer made it look like a person was riding on its back. A grasshopper is not even a mammal, and it has an exoskeleton instead of an internal skeleton. The whole argument would be a lot more convincing if we could do some isolation of variables, a scientific term that means to change only one thing at a time, isolating it from the other variables that might have an effect. If size is the variable whose effect we're

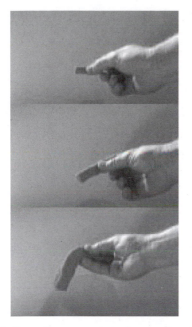

Galileo discusses planks made of wood, but the concept may be easier to imagine with clay. All three clay rods in the figure were originally the same shape. The medium-sized one was twice the height, twice the length, and twice the width of the small one, and similarly the large one was twice as big as the medium one in all its linear dimensions. The big one has four times the linear dimensions of the small one, 16 times the cross-sectional area when cut perpendicular to the page, and 64 times the volume. That means that the big one has 64 times the weight to support, but only 16 times the strength compared to the smallest one.

interested in seeing, then we don't really want to compare things that are different in size but also different in other ways.

Also, Galileo is doing something that would be frowned on in modern science: he is mixing experiments whose results he has actually observed (building boats of different sizes), with experiments that he could not possibly have done (dropping an ant from the height of the moon).

After this entertaining but not scientifically rigorous beginning, Galileo starts to do something worthwhile by modern standards. He simplifies everything by considering the strength of a wooden plank. The variables involved can then be narrowed down to the type of wood, the width, the thickness, and the length. He also gives an operational definition of what it means for the plank to have a certain strength "in proportion to its size," by introducing the concept of a plank that is the longest one that would not snap under its own weight if supported at one end. If you increased its length by the slightest amount, without increasing its width or thickness, it would break. He says that if one plank is the same shape as another but a different size, appearing like a reduced or enlarged photograph of the other, then the planks would be strong "in proportion to their sizes" if both were just barely able to support their own weight.

He now relates how he has done actual experiments with such planks, and found that, according to this operational definition, they are not strong in proportion to their sizes. The larger one breaks. He makes sure to tell the reader how important the result is, via Sagredo's astonished response:

SAGREDO: My brain already reels. My mind, like a cloud momentarily illuminated by a lightning flash, is for an instant filled with an unusual light, which now beckons to me and which now suddenly mingles and obscures strange, crude ideas. From what you have said it appears to me impossible to build two similar structures of the same material, but of different sizes and have them proportionately strong.

In other words, this specific experiment, using things like wooden planks that have no intrinsic scientific interest, has very wide implications because it points out a general principle, that nature acts differently on different scales.

To finish the discussion, Galileo gives an explanation. He says that the strength of a plank (defined as, say, the weight of the heaviest boulder you could put on the end without breaking it) is proportional to its cross-sectional area, that is, the surface area of the fresh wood that would be exposed if you sawed through it in the middle. Its weight, however, is proportional to its volume.*

How do the volume and cross-sectional area of the longer plank compare with those of the shorter plank? We have already seen, while discussing conversions of the units of area and volume, that these quantities don't act the way most people naively expect. You might think that the volume and area of the longer plank would both be doubled compared to the shorter plank, so they would increase in proportion to each other, and the longer plank would be equally able to support its weight. You would be wrong, but Galileo knows that this is a common misconception, so he has

---

*Galileo makes a slightly more complicated argument, taking into account the effect of leverage (torque). The result I'm referring to comes out the same regardless of this effect.

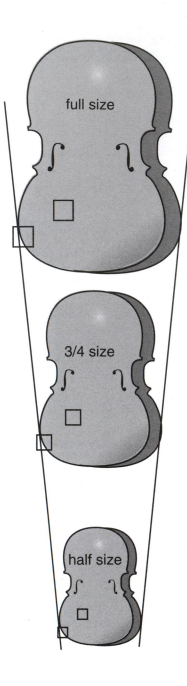

full size

3/4 size

half size

Salviati address the point specifically:

SALVIATI: ...Take, for example, a cube two inches on a side so that each face has an area of four square inches and the total area, i.e., the sum of the six faces, amounts to twenty-four square inches; now imagine this cube to be sawed through three times [with cuts in three perpendicular planes] so as to divide it into eight smaller cubes, each one inch on the side, each face one inch square, and the total surface of each cube six square inches instead of twenty-four in the case of the larger cube. It is evident therefore, that the surface of the little cube is only one-fourth that of the larger, namely, the ratio of six to twenty-four; but the volume of the solid cube itself is only one-eighth; the volume, and hence also the weight, diminishes therefore much more rapidly than the surface... You see, therefore, Simplicio, that I was not mistaken when ... I said that the surface of a small solid is comparatively greater than that of a large one.

The same reasoning applies to the planks. Even though they are not cubes, the large one could be sawed into eight small ones, each with half the length, half the thickness, and half the width. The small plank, therefore, has more surface area in proportion to its weight, and is therefore able to support its own weight while the large one breaks.

Scaling of area and volume for irregularly shaped objects

You probably are not going to believe Galileo's claim that this has deep implications for all of nature unless you can be convinced that the same is true for any shape. Every drawing you've seen so far has been of squares, rectangles, and rectangular solids. Clearly the reasoning about sawing things up into smaller pieces would not prove anything about, say, an egg, which cannot be cut up into eight smaller egg-shaped objects with half the length.

Is it always true that something half the size has one quarter the surface area and one eighth the volume, even if it has an irregular shape? Take the example of a child's violin. Violins are made for small children in lengths that are either half or 3/4 of the normal length, accommodating their small hands. Let's study the surface area of the front panels of the three violins.

Consider the square in the interior of the panel of the full-size violin. In the 3/4-size violin, its height and width are both smaller by a factor of 3/4, so the area of the corresponding, smaller square becomes 3/4x3/4=9/16 of the original area, not 3/4 of the original area. Similarly, the corresponding square on the smallest violin has half the height and half the width of the original one, so its area is 1/4 the original area, not half.

The same reasoning works for parts of the panel near the edge, such as the part that only partially fills in the other square. The entire square scales down the same as a square in the interior, and in each violin the same fraction (about 70%) of the square is full, so the contribution of this part to the total area scales down just the same.

Since any small square region or any small region covering part of a square scales down like a square object, the entire surface area of an irregularly shaped object changes in the same manner as the surface area of a square: scaling it down by 3/4 reduces the area by a factor of 9/16, and so on.

In general, we can see that any time there are two objects with the same shape, but different linear dimensions (i.e. one looks like a reduced photo of the other), the ratio of their areas equals the ratio of the squares of their linear dimensions:

$$\frac{A_1}{A_2} = \left(\frac{L_1}{L_2}\right)^2 \quad .$$

Note that it doesn't matter where we choose to measure the linear size, L, of an object. In the case of the violins, for instance, it could have been measured vertically, horizontally, diagonally, or even from the bottom of the left f-hole to the middle of the right f-hole. We just have to measure it in a consistent way on each violin. Since all the parts are assumed to shrink or expand in the same manner, the ratio L_1/L_2 is independent of the choice of measurement.

It is also important to realize that it is completely unnecessary to have a formula for the area of a violin. It is only possible to derive simple formulas for the areas of certain shapes like circles, rectangles, triangles and so on, but that is no impediment to the type of reasoning we are using.

Sometimes it is inconvenient to write all the equations in terms of ratios, especially when more than two objects are being compared. A more compact way of rewriting the previous equation is

$$A \propto L^2 \quad .$$

The symbol "$\propto$" means "is proportional to." Scientists and engineers often speak about such relationships verbally using the phrases "scales like" or "goes like," for instance "area goes like length squared."

All of the above reasoning works just as well in the case of volume. Volume goes like length cubed:

$$V \propto L^3 \quad .$$

If different objects are made of the same material with the same density, $\rho = m/V$, then their masses, $m = \rho V$, are proportional to L^3, and so are their weights. (The symbol for density is ρ, the lower-case Greek letter "rho".)

An important point is that all of the above reasoning about scaling only applies to objects that are the same shape. For instance, a piece of paper is larger than a pencil, but has a much greater surface-to-volume ratio.

One of the first things I learned as a teacher was that students were not very original about their mistakes. Every group of students tends to come up with the same goofs as the previous class. The following are some examples of correct and incorrect reasoning about proportionality.

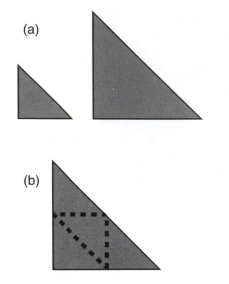

(a)

(b)

The big triangle has four times more area than the little one.

Example: scaling of the area of a triangle

Question: In fig. (a), the larger triangle has sides twice as long. How many times greater is its area?

Correct solution #1: Area scales in proportion to the square of the linear dimensions, so the larger triangle has four times more area (2^2=4).

Correct solution #2: You could cut the larger triangle into four of the smaller size, as shown in fig. (b), so its area is four times greater. (This solution is correct, but it would not work for a shape like a circle, which can't be cut up into smaller circles.)

Correct solution #3: The area of a triangle is given by

$A = \frac{1}{2} bh$, where b is the base and h is the height. The areas of the

triangles are

$$A_1 = \frac{1}{2} b_1 h_1$$
$$A_2 = \frac{1}{2} b_2 h_2$$
$$= \frac{1}{2} (2b_1)(2h_1)$$
$$= 2b_1 h_1$$
$$A_2/A_1 = (2b_1 h_1)/(\frac{1}{2} b_1 h_1)$$
$$= 4$$

(Although this solution is correct, it is a lot more work than solution #1, and it can only be used in this case because a triangle is a simple geometric shape, and we happen to know a formula for its area.)

Correct solution #4: The area of a triangle is $A = \frac{1}{2} bh$. The

comparison of the areas will come out the same as long as the ratios of the linear sizes of the triangles is as specified, so let's just say b_1=1.00 m and b_2=2.00 m. The heights are then also h_1=1.00 m and h_2=2.00 m, giving areas A_1=0.50 m^2 and A_2=2.00 m^2, so A_2/A_1=4.00.

(The solution is correct, but it wouldn't work with a shape for whose area we don't have a formula. Also, the numerical calculation might make the answer of 4.00 appear inexact, whereas solution #1 makes it clear that it is exactly 4.)

Incorrect solution: The area of a triangle is $A = \frac{1}{2} bh$, and if you

plug in b=2.00 m and h=2.00 m, you get A=2.00 m^2, so the bigger triangle has 2.00 times more area. (This solution is incorrect because no comparison has been made with the smaller triangle.)

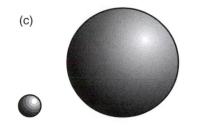

(c)

The big sphere has 125 times more volume than the little one.

Example: scaling of the volume of a sphere
Question: In figure (c), the larger sphere has a radius that is five times greater. How many times greater is its volume?
Correct solution #1: Volume scales like the third power of the linear size, so the larger sphere has a volume that is 125 times greater (5^3=125).

Correct solution #2: The volume of a sphere is $V=\frac{4}{3}\pi r^3$, so

$$V_1 = \frac{4}{3}\pi r_1^3$$

$$V_2 = \frac{4}{3}\pi r_2^3$$

$$= \frac{4}{3}\pi(5r_1)^3$$

$$= \frac{500}{3}\pi r_1^3$$

$$V_2/V_1 = \left(\frac{500}{3}\pi r_1^3\right) / \left(\frac{4}{3}\pi r_1^3\right)$$

$$= 125$$

Incorrect solution: The volume of a sphere is $V=\frac{4}{3}\pi r^3$, so

$$V_1 = \frac{4}{3}\pi r_1^3$$

$$V_2 = \frac{4}{3}\pi r_2^3$$

$$= \frac{4}{3}\pi \cdot 5r_1^3$$

$$= \frac{20}{3}\pi r_1^3$$

$$V_2/V_1 = (\frac{20}{3}\pi r_1^3)/(\frac{4}{3}\pi r_1^3)$$

$$= 5$$

(The solution is incorrect because $(5r_1)^3$ is not the same as $5r_1^3$.)

S S

(d) The 48-point "S" has 1.78 times more area than the 36-point "S."

Example: scaling of a more complex shape
Question: The first letter "S" in fig. (d) is in a 36-point font, the second in 48-point. How many times more ink is required to make the larger "S"?
Correct solution: The amount of ink depends on the area to be covered with ink, and area is proportional to the square of the linear dimensions, so the amount of ink required for the second "S" is greater by a factor of $(48/36)^2$=1.78.
Incorrect solution: The length of the curve of the second "S" is longer by a factor of 48/36=1.33, so 1.33 times more ink is required.
(The solution is wrong because it assumes incorrectly that the width of the curve is the same in both cases. Actually both the width and the length of the curve are greater by a factor of 48/36, so the area is greater by a factor of $(48/36)^2$=1.78.)

Discussion questions

A. A toy fire engine is 1/30 the size of the real one, but is constructed from the same metal with the same proportions. How many times smaller is its weight? How many times less red paint would be needed to paint it?

B. Galileo spends a lot of time in his dialog discussing what really happens when things break. He discusses everything in terms of Aristotle's now-discredited explanation that things are hard to break, because if something breaks, there has to be a gap between the two halves with nothing in between, at least initially. Nature, according to Aristotle, "abhors a vacuum," i.e. nature doesn't "like" empty space to exist. Of course, air will rush into the gap immediately, but at the very moment of breaking, Aristotle imagined a vacuum in the gap. Is Aristotle's explanation of why it is hard to break things an experimentally testable statement? If so, how could it be tested experimentally?

1.3 Scaling Applied to Biology

Organisms of different sizes with the same shape

The first of the following graphs shows the approximate validity of the proportionality $m \propto L^3$ for cockroaches (redrawn from McMahon and Bonner). The scatter of the points around the curve indicates that some cockroaches are proportioned slightly differently from others, but in general the data seem well described by $m \propto L^3$. That means that the largest cockroaches the experimenter could raise (is there a 4-H prize?) had roughly the same shape as the smallest ones.

Another relationship that should exist for animals of different sizes shaped in the same way is that between surface area and body mass. If all the animals have the same average density, then body mass should be proportional to the cube of the animal's linear size, $m \propto L^3$, while surface area should vary proportionately to L^2. Therefore, the animals' surface areas should be proportional to $m^{2/3}$. As shown in the second graph, this relationship appears to hold quite well for the dwarf siren, a type of salamander. Notice how the curve bends over, meaning that the surface area does not increase as quickly as body mass, e.g. a salamander with eight times more body mass will have only four times more surface area.

This behavior of the ratio of surface area to mass (or, equivalently, the ratio of surface area to volume) has important consequences for mammals, which must maintain a constant body temperature. It would make sense for the rate of heat loss through the animal's skin to be proportional to its surface area, so we should expect small animals, having large ratios of surface area to volume, to need to produce a great deal of heat in comparison to their size to avoid dying from low body temperature. This expectation is borne out by the data of the third graph, showing the rate of oxygen consumption of guinea pigs as a function of their body mass. Neither an animal's heat production nor its surface area is convenient to measure, but in order to produce heat, the animal must metabolize oxygen, so oxygen consumption is a good indicator of the rate of heat production. Since surface area is proportional to $m^{2/3}$, the proportionality of the rate of oxygen consumption to $m^{2/3}$ is consistent with the idea that the animal needs to produce heat at a rate in proportion to its surface area. Although the smaller animals metabolize less oxygen and produce less heat in absolute terms, the amount of food and oxygen they must consume is greater in proportion to their own mass. The Etruscan pigmy shrew, weighing in at 2 grams as an

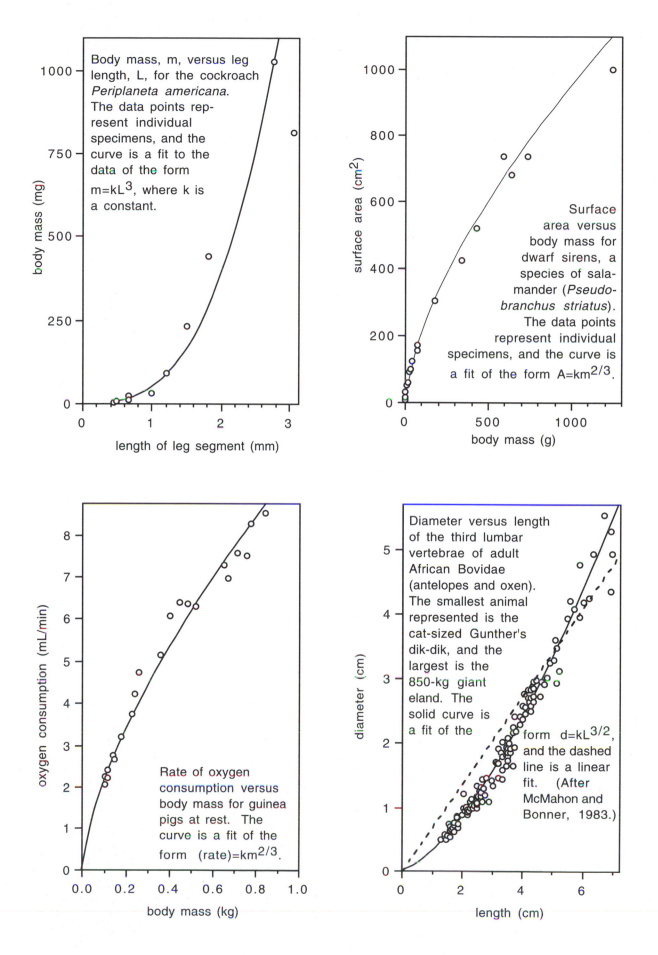

Body mass, m, versus leg length, L, for the cockroach *Periplaneta americana*. The data points represent individual specimens, and the curve is a fit to the data of the form $m=kL^3$, where k is a constant.

Surface area versus body mass for dwarf sirens, a species of salamander (*Pseudobranchus striatus*). The data points represent individual specimens, and the curve is a fit of the form $A=km^{2/3}$.

Rate of oxygen consumption versus body mass for guinea pigs at rest. The curve is a fit of the form $(rate)=km^{2/3}$.

Diameter versus length of the third lumbar vertebrae of adult African Bovidae (antelopes and oxen). The smallest animal represented is the cat-sized Gunther's dik-dik, and the largest is the 850-kg giant eland. The solid curve is a fit of the form $d=kL^{3/2}$, and the dashed line is a linear fit. (After McMahon and Bonner, 1983.)

adult, is at about the lower size limit for mammals. It must eat continually, consuming many times its body weight each day to survive.

Changes in shape to accommodate changes in size

Large mammals, such as elephants, have a small ratio of surface area to volume, and have problems getting rid of their heat fast enough. An elephant cannot simply eat small enough amounts to keep from producing excessive heat, because cells need to have a certain minimum metabolic rate to run their internal machinery. Hence the elephant's large ears, which add to its surface area and help it to cool itself. Previously, we have seen several examples of data within a given species that were consistent with a fixed shape, scaled up and down in the cases of individual specimens. The elephant's ears are an example of a change in shape necessitated by a change in scale.

Large animals also must be able to support their own weight. Returning to the example of the strengths of planks of different sizes, we can see that if the strength of the plank depends on area while its weight depends on volume, then the ratio of strength to weight goes as follows:

$$\text{strength/weight} \propto A/V \propto 1/L \quad .$$

Thus, the ability of objects to support their own weights decreases inversely in proportion to their linear dimensions. If an object is to be just barely able to support its own weight, then a larger version will have to be proportioned differently, with a different shape.

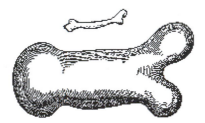

Galileo's original drawing, showing how larger animals' bones must be greater in diameter compared to their lengths.

Since the data on the cockroaches seemed to be consistent with roughly similar shapes within the species, it appears that the ability to support its own weight was not the tightest design constraint that Nature was working under when she designed them. For large animals, structural strength is important. Galileo was the first to quantify this reasoning and to explain why, for instance, a large animal must have bones that are thicker in proportion to their length. Consider a roughly cylindrical bone such as a leg bone or a vertebra. The length of the bone, L, is dictated by the overall linear size of the animal, since the animal's skeleton must reach the animal's whole length. We expect the animal's mass to scale as L^3, so the strength of the bone must also scale as L^3. Strength is proportional to cross-sectional area, as with the wooden planks, so if the diameter of the bone is d, then

$$d^2 \propto L^3$$

or

$$d \propto L^{3/2} \quad .$$

If the shape stayed the same regardless of size, then all linear dimensions, including d and L, would be proportional to one another. If our reasoning holds, then the fact that d is proportional to $L^{3/2}$, not L, implies a change in proportions of the bone. As shown in the graph on the previous page, the vertebrae of African Bovidae follow the rule $d \propto L^{3/2}$ fairly well. The vertebrae of the giant eland are as chunky as a coffee mug, while those of a Gunther's dik-dik are as slender as the cap of a pen.

A. Single-celled animals must passively absorb nutrients and oxygen from their surroundings, unlike humans who have lungs to pump air in and out and a heart to distribute the oxygenated blood throughout their bodies. Even the cells composing the bodies of multicellular animals must absorb oxygen from a nearby capillary through their surfaces. Based on these facts, explain why cells are always microscopic in size.

B. The reasoning of the previous question would seem to be contradicted by the fact that human nerve cells in the spinal cord can be as much as a meter long, although their widths are still very small. Why is this possible?

1.4   Order-of-Magnitude Estimates

It is the mark of an instructed mind to rest satisfied with the degree of precision that the nature of the subject permits and not to seek an exactness where only an approximation of the truth is possible.

Aristotle

It is a common misconception that science must be exact. For instance, in the Star Trek TV series, it would often happen that Captain Kirk would ask Mr. Spock, "Spock, we're in a pretty bad situation. What do you think are our chances of getting out of here?" The scientific Mr. Spock would answer with something like, "Captain, I estimate the odds as 237.345 to one." In reality, he could not have estimated the odds with six significant figures of accuracy, but nevertheless one of the hallmarks of a person with a good education in science is the ability to make estimates that are likely to be at least somewhere in the right ballpark. In many such situations, it is often only necessary to get an answer that is off by no more than a factor of ten in either direction. Since things that differ by a factor of ten are said to differ by one order of magnitude, such an estimate is called an order-of-magnitude estimate. The tilde, ~, is used to indicate that things are only of the same order of magnitude, but not exactly equal, as in

odds of survival ~ 100 to one .

The tilde can also be used in front of an individual number to emphasize that the number is only of the right order of magnitude.

Although making order-of-magnitude estimates seems simple and natural to experienced scientists, it's a mode of reasoning that is completely unfamiliar to most college students. Some of the typical mental steps can be illustrated in the following example.

Example: Cost of transporting tomatoes
Question: Roughly what percentage of the price of a tomato comes from the cost of transporting it in a truck?

The following incorrect solution illustrates one of the main ways you can go wrong in order-of-magnitude estimates.

> **Incorrect solution:** Let's say the trucker needs to make a $400 profit on the trip. Taking into account her benefits, the cost of gas, and maintenance and payments on the truck, let's say the total cost is more like $2000. I'd guess about 5000 tomatoes would fit in the back of the truck, so the extra cost per tomato is 40 cents. That means the cost of transporting one tomato is comparable to the cost of the tomato itself. Transportation really adds a lot to the cost of produce, I guess.

The problem is that the human brain is not very good at estimating area or volume, so it turns out the estimate of 5000 tomatoes fitting in the truck is way off. That's why people have a hard time at those contests where you are supposed to estimate the number of jellybeans in a big jar. Another example is that most people think their families use about 10 gallons of water per day, but in reality the average is about 300 gallons per day. When estimating area or volume, you are much better off estimating linear dimensions, and computing volume from the linear dimensions. Here's a better solution:

> **Better solution:** As in the previous solution, say the cost of the trip is $2000. The dimensions of the bin are probably 4 m x 2 m x 1 m, for a volume of 8 m^3. Since the whole thing is just an order-of-magnitude estimate, let's round that off to the nearest power of ten, 10 m^3. The shape of a tomato is complicated, and I don't know any formula for the volume of a tomato shape, but since this is just an estimate, let's pretend that a tomato is a cube, 0.05 m x 0.05 m x 0.05, for a volume of 1.25×10^{-4} m^3. Since this is just a rough estimate, let's round that to 10^{-4} m^3. We can find the total number of tomatoes by dividing the volume of the bin by the volume of one tomato: 10 m^3 / 10^{-4} m^3 = 10^5 tomatoes. The transportation cost per tomato is $2000/$10^5$ tomatoes=$0.02/tomato. That means that transportation really doesn't contribute very much to the cost of a tomato.

Approximating the shape of a tomato as a cube is an example of another general strategy for making order-of-magnitude estimates. A similar situation would occur if you were trying to estimate how many m^2 of leather could be produced from a herd of ten thousand cattle. There is no point in trying to take into account the shape of the cows' bodies. A reasonable plan of attack might be to consider a spherical cow. Probably a cow has roughly the same surface area as a sphere with a radius of about 1 m, which would be $4\pi(1 \text{ m})^2$. Using the well-known facts that pi equals three, and four times three equals about ten, we can guess that a cow has a surface area of about 10 m^2, so the herd as a whole might yield 10^5 m^2 of leather.

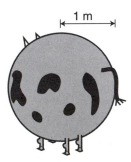

1 m

The following list summarizes the strategies for getting a good order-of-magnitude estimate.

(1) Don't even attempt more than one significant figure of precision.

(2) Don't guess area or volume directly. Guess linear dimensions and get area or volume from them.

(3) When dealing with areas or volumes of objects with complex shapes, idealize them as if they were some simpler shape, a cube or a sphere, for example.

(4) Check your final answer to see if it is reasonable. If you estimate that a herd of ten thousand cattle would yield 0.01 m^2 of leather, then you have probably made a mistake with conversion factors somewhere.

Summary

Notation

∝ .... is proportional to

~ .... on the order of, is on the order of

Summary

Nature behaves differently on large and small scales. Galileo showed that this results fundamentally from the way area and volume scale. Area scales as the second power of length, $A \propto L^2$, while volume scales as length to the third power, $V \propto L^3$.

An order of magnitude estimate is one in which we do not attempt or expect an exact answer. The main reason why the uninitiated have trouble with order-of-magnitude estimates is that the human brain does not intuitively make accurate estimates of area and volume. Estimates of area and volume should be approached by first estimating linear dimensions, which one's brain has a feel for.

Homework Problems

1 ✓. How many cubic inches are there in a cubic foot? The answer is not 12.

2. Assume a dog's brain is twice is great in diameter as a cat's, but each animal's brain cells are the same size and their brains are the same shape. In addition to being a far better companion and much nicer to come home to, how many times more brain cells does a dog have than a cat? The answer is not 2.

3 ✓. The population density of Los Angeles is about 4000 people/km². That of San Francisco is about 6000 people/km². How many times farther away is the average person's nearest neighbor in LA than in San Francisco? The answer is not 1.5.

4. A hunting dog's nose has about 10 square inches of active surface. How is this possible, since the dog's nose is only about 1 in x 1 in x 1 in = 1 in³? After all, 10 is greater than 1, so how can it fit?

5. Estimate the number of blades of grass on a football field.

6. In a computer memory chip, each bit of information (a 0 or a 1) is stored in a single tiny circuit etched onto the surface of a silicon chip. The circuits cover the surface of the chip like lots in a housing development. A typical chip stores 64 Mb (megabytes) of data, where a byte is 8 bits. Estimate (a) the area of each circuit, and (b) its linear size.

7. Suppose someone built a gigantic apartment building, measuring 10 km x 10 km at the base. Estimate how tall the building would have to be to have space in it for the entire world's population to live.

8. A hamburger chain advertises that it has sold 10 billion Bongo Burgers. Estimate the total mass of feed required to raise the cows used to make the burgers.

9. Estimate the volume of a human body, in cm³.

10 S. How many cm² is 1 mm²?

S A solution is given in the back of the book.	★ A difficult problem.
✓ A computerized answer check is available.	∫ A problem that requires calculus.

11 S. Compare the light-gathering powers of a 3-cm-diameter telescope and a 30-cm telescope.

12. S. One step on the Richter scale corresponds to a factor of 100 in terms of the energy absorbed by something on the surface of the Earth, e.g. a house. For instance, a 9.3-magnitude quake would release 100 times more energy than an 8.3. The energy spreads out from the epicenter as a wave, and for the sake of this problem we'll assume we're dealing with seismic waves that spread out in three dimensions, so that we can visualize them as hemispheres spreading out under the surface of the earth. If a certain 7.6-magnitude earthquake and a certain 5.6-magnitude earthquake produce the same amount of vibration where I live, compare the distances from my house to the two epicenters.

13✓. In Europe, a piece of paper of the standard size, called A4, is a little narrower and taller than its American counterpart. The ratio of the height to the width is the square root of 2, and this has some useful properties. For instance, if you cut an A4 sheet from left to right, you get two smaller sheets that have the same proportions. You can even buy sheets of this smaller size, and they're called A5. There is a whole series of sizes related in this way, all with the same proportions. (a) Compare an A5 sheet to an A4 in terms of area and linear size. (b) The series of paper sizes starts from an A0 sheet, which has an area of one square meter. Suppose we had a series of boxes defined in a similar way: the B0 box has a volume of one cubic meter, two B1 boxes fit exactly inside an B0 box, and so on. What would be the dimensions of a B0 box?

14. Estimate the mass of a human hair, in units of kg.

Motion in One Dimension

I didn't learn until I was nearly through with college that I could understand a book much better if I mentally outlined it for myself before I actually began reading. It's a technique that warns my brain to get little cerebral file folders ready for the different topics I'm going to learn, and as I'm reading it allows me to say to myself, "Oh, the reason they're talking about this now is because they're preparing for this other thing that comes later," or "I don't need to sweat the details of this idea now, because they're going to explain it in more detail later on."

At this point, you're about to dive in to the main subjects of this book, which are force and motion. The concepts you're going to learn break down into the following three areas:

kinematics — how to describe motion numerically
dynamics — how force affects motion
vectors — a mathematical way of handling the three-dimensional nature of force and motion

Roughly speaking, that's the order in which we'll cover these three areas, but the earlier chapters do contain quite a bit of preparation for the later topics. For instance, even before the present point in the book you've learned about the Newton, a unit of force. The discussion of force properly belongs to dynamics, which we aren't tackling head-on for a few more chapters, but I've found that when I teach kinematics it helps to be able to refer to forces now and then to show why it makes sense to define certain kinematical concepts. And although I don't explicitly introduce vectors until ch. 8, the groundwork is being laid for them in earlier chapters.

Here's a roadmap to the rest of the book:

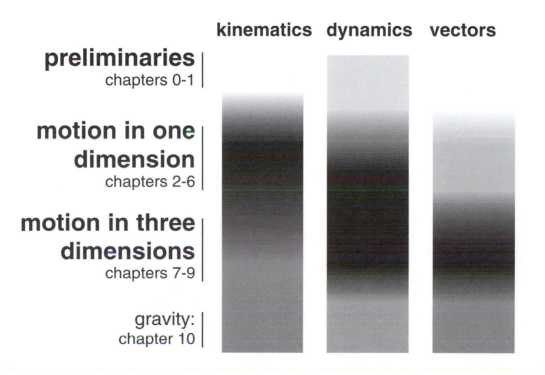

2 Velocity and Relative Motion

2.1 Types of Motion

Rotation.

Simultaneous rotation and motion through space.

One person might say that the tipping chair was only rotating in a circle about its point of contact with the floor, but another could describe it as having both rotation and motion through space.

If you had to think consciously in order to move your body, you would be severely disabled. Even walking, which we consider to be no great feat, requires an intricate series of motions that your cerebrum would be utterly incapable of coordinating. The task of putting one foot in front of the other is controlled by the more primitive parts of your brain, the ones that have not changed much since the mammals and reptiles went their separate evolutionary ways. The thinking part of your brain limits itself to general directives such as "walk faster," or "don't step on her toes," rather than micromanaging every contraction and relaxation of the hundred or so muscles of your hips, legs, and feet.

Physics is all about the conscious understanding of motion, but we're obviously not immediately prepared to understand the most complicated types of motion. Instead, we'll use the divide-and-conquer technique. We'll first classify the various types of motion, and then begin our campaign with an attack on the simplest cases. To make it clear what we are and are not ready to consider, we need to examine and define carefully what types of motion can exist.

Rigid-body motion distinguished from motion that changes an object's shape

Nobody, with the possible exception of Fred Astaire, can simply glide forward without bending their joints. Walking is thus an example in which there is both a general motion of the whole object and a change in the shape of the object. Another example is the motion of a jiggling water balloon as it flies through the air. We are not presently attempting a mathematical description of the way in which the shape of an object changes. Motion without a change in shape is called rigid-body motion. (The word "body" is often used in physics as a synonym for "object.")

Center-of-mass motion as opposed to rotation

A ballerina leaps into the air and spins around once before landing. We feel intuitively that her rigid-body motion while her feet are off the ground consists of two kinds of motion going on simultaneously: a rotation and a motion of her body as a whole through space, along an arc. It is not immediately obvious, however, what is the most useful way to define the distinction between rotation and motion through space. Imagine that you attempt to balance a chair and it falls over. One person might say that the only motion was a rotation about the chair's point of contact with the floor, but another might say that there was both rotation and motion down and to the side.

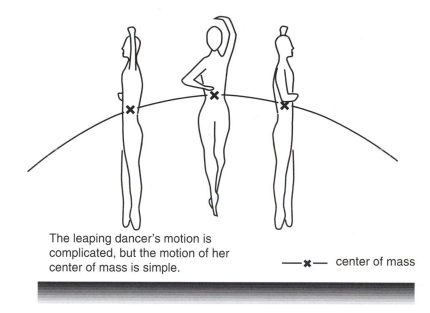

The leaping dancer's motion is complicated, but the motion of her center of mass is simple.

——✗—— center of mass

No matter what point you hang the pear from, the string lines up with the pear's center of mass. The center of mass can therefore be defined as the intersection of all the lines made by hanging the pear in this way. Note that the X in the figure should not be interpreted as implying that the center of mass is on the surface — it is actually inside the pear.

It turns out that there is one particularly natural and useful way to make a clear definition, but it requires a brief digression. Every object has a balance point, referred to in physics as the *center of mass*. For a two-dimensional object such as a cardboard cutout, the center of mass is the point at which you could hang the object from a string and make it balance. In the case of the ballerina (who is likely to be three-dimensional unless her diet is particularly severe), it might be a point either inside or outside her body, depending on how she holds her arms. Even if it is not practical to attach a string to the balance point itself, the center of mass can be defined as shown in the figure on the left.

Why is the center of mass concept relevant to the question of classifying rotational motion as opposed to motion through space? As illustrated in the figure above, it turns out that the motion of an object's center of mass is nearly always far simpler than the motion of any other part of the object. The ballerina's body is a large object with a complex shape. We might expect that her motion would be much more complicated that the motion of a small, simply-shaped object, say a marble, thrown up at the same angle as the angle at which she leapt. But it turns out that the motion of the ballerina's center of mass is exactly the same as the motion of the marble. That is, the motion of the center of mass is the same as the motion the ballerina would have if all her mass was concentrated at a point. By restricting our attention to the motion of the center of mass, we can therefore simplify things greatly.

The same leaping dancer, viewed from above. Her center of mass traces a straight line, but a point away from her center of mass, such as her elbow, traces the much more complicated path shown by the dots.

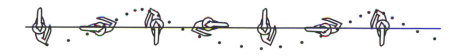

We can now replace the ambiguous idea of "motion as a whole through space" with the more useful and better defined concept of *"center-of-mass motion."* The motion of any rigid body can be cleanly split into rotation and center-of-mass motion. By this definition, the tipping chair does have both rotational and center-of-mass motion. Concentrating on the center of

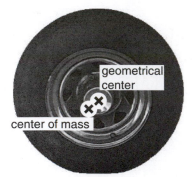

An improperly balanced wheel has a center of mass that is not at its geometric center. When you get a new tire, the mechanic clamps little weights to the rim to balance the wheel.

mass motion allows us to make a simplified *model* of the motion, as if a complicated object like a human body was just a marble or a point-like particle. Science really never deals with reality; it deals with models of reality.

Note that the word "center" in "center of mass" is not meant to imply that the center of mass must lie at the geometrical center of an object. A car wheel that has not been balanced properly has a center of mass that does not coincide with its geometrical center. An object such as the human body does not even have an obvious geometrical center.

It can be helpful to think of the center of mass as the average location of all the mass in the object. With this interpretation, we can see for example that raising your arms above your head raises your center of mass, since the

A fixed point on the dancer's body follows a trajectory that is flatter than what we expect, creating an illusion of flight.

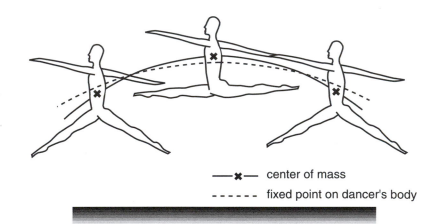

———✕——— center of mass

- - - - - fixed point on dancer's body

higher position of the arms' mass raises the average.

Ballerinas and professional basketball players can create an illusion of flying horizontally through the air because our brains intuitively expect them to have rigid-body motion, but the body does not stay rigid while executing a grand jete or a slam dunk. The legs are low at the beginning and end of the jump, but come up higher at the middle. Regardless of what the limbs do, the center of mass will follow the same arc, but the low position of the legs at the beginning and end means that the torso is higher compared to the center of mass, while in the middle of the jump it is lower compared to the center of mass. Our eye follows the motion of the torso and tries to interpret it as the center-of-mass motion of a rigid body. But since the torso follows a path that is flatter than we expect, this attempted interpretation fails, and we experience an illusion that the person is flying horizontally. Another interesting example from the sports world is the high jump, in which the jumper's curved body passes over the bar, but the center of mass passes under the bar! Here the jumper lowers his legs and upper body at the peak of the jump in order to bring his waist higher compared to the center of mass.

The high-jumper's body passes over the bar, but his center of mass passes under it.
Photo by Dunia Young.

Later in this course, we'll find that there are more fundamental reasons (based on Newton's laws of motion) why the center of mass behaves in such a simple way compared to the other parts of an object. We're also postponing any discussion of numerical methods for finding an object's center of mass. Until later in the course, we will only deal with the motion of objects' centers of mass.

Center-of-mass motion in one dimension

In addition to restricting our study of motion to center-of-mass motion, we will begin by considering only cases in which the center of mass moves along a straight line. This will include cases such as objects falling straight down, or a car that speeds up and slows down but does not turn.

Note that even though we are not explicitly studying the more complex aspects of motion, we can still analyze the center-of-mass motion while ignoring other types of motion that might be occurring simultaneously. For instance, if a cat is falling out of a tree and is initially upside-down, it goes through a series of contortions that bring its feet under it. This is definitely not an example of rigid-body motion, but we can still analyze the motion of the cat's center of mass just as we would for a dropping rock.

Self-Check

Consider a person running, a person pedaling on a bicycle, a person coasting on a bicycle, and a person coasting on ice skates. In which cases is the center-of-mass motion one-dimensional? Which cases are examples of rigid-body motion?

2.2   Describing Distance and Time

Center-of-mass motion in one dimension is particularly easy to deal with because all the information about it can be encapsulated in two variables: x, the position of the center of mass relative to the origin, and t, which measures a point in time. For instance, if someone supplied you with a sufficiently detailed table of x and t values, you would know pretty much all there was to know about the motion of the object's center of mass.

A point in time as opposed to duration

In ordinary speech, we use the word "time" in two different senses, which are to be distinguished in physics. It can be used, as in "a short time" or "our time here on earth," to mean a length or duration of time, or it can be used to indicate a clock reading, as in "I didn't know what time it was," or "now's the time." In symbols, t is ordinarily used to mean a point in time, while Δt signifies an interval or duration in time. The capital Greek letter delta, Δ, means "the change in...," i.e. a duration in time is the change or difference between one clock reading and another. The notation Δt does not signify the product of two numbers, Δ and t, but rather one single number, Δt. If a matinee begins at a point in time $t=1$ o'clock and ends at $t=3$ o'clock, the duration of the movie was the change in t,

$$\Delta t = 3 \text{ hours} - 1 \text{ hour} = 2 \text{ hours} \quad .$$

To avoid the use of negative numbers for Δt, we write the clock reading "after" to the left of the minus sign, and the clock reading "before" to the right of the minus sign. A more specific definition of the delta notation is therefore that delta stands for "after minus before."

Even though our definition of the delta notation guarantees that Δt is positive, there is no reason why t can't be negative. If t could not be negative, what would have happened one second before $t=0$? That doesn't mean

Coasting on a bike and coasting on skates give one-dimensional center-of-mass motion, but running and pedaling require moving body parts up and down, which makes the center of mass move up and down. The only example of rigid-body motion is coasting on skates. (Coasting on a bike is not rigid-body motion, because the wheels twist.)

that time "goes backward" in the sense that adults can shrink into infants and retreat into the womb. It just means that we have to pick a reference point and call it $t=0$, and then times before that are represented by negative values of t.

Although a point in time can be thought of as a clock reading, it is usually a good idea to avoid doing computations with expressions such as "2:35" that are combinations of hours and minutes. Times can instead be expressed entirely in terms of a single unit, such as hours. Fractions of an hour can be represented by decimals rather than minutes, and similarly if a problem is being worked in terms of minutes, decimals can be used instead of seconds.

Self-Check

Of the following phrases, which refer to points in time, which refer to time intervals, and which refer to time in the abstract rather than as a measurable number?

(a) "The time has come."
(b) "Time waits for no man."
(c) "The whole time, he had spit on his chin."

Position as opposed to change in position

As with time, a distinction should be made between a point in space, symbolized as a coordinate x, and a change in position, symbolized as Δx.

As with t, x can be negative. If a train is moving down the tracks, not only do you have the freedom to choose any point along the tracks and call it $x=0$, but it's also up to you to decide which side of the $x=0$ point is positive x and which side is negative x.

Since we've defined the delta notation to mean "after minus before," it is possible that Δx will be negative, unlike Δt which is guaranteed to be positive. Suppose we are describing the motion of a train on tracks linking Tucson and Chicago. As shown in the figure, it is entirely up to you to decide which way is positive.

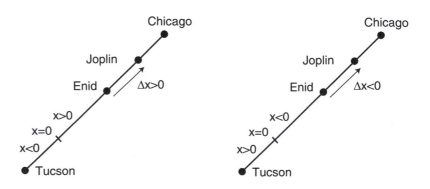

Two equally valid ways of describing the motion of a train from Tucson to Chicago. In the first example, the train has a positive Δx as it goes from Enid to Joplin. In the second example, the same train going forward in the same direction has a negative Δx.

(a) a point in time; (b) time in the abstract sense; (c) a time interval

Note that in addition to x and Δx, there is a third quantity we could define, which would be like an odometer reading, or actual distance traveled. If you drive 10 miles, make a U-turn, and drive back 10 miles, then your Δx is zero, but your car's odometer reading has increased by 20 miles. However important the odometer reading is to car owners and used car dealers, it is not very important in physics, and there is not even a standard name or notation for it. The change in position, Δx, is more useful because it is so much easier to calculate: to compute Δx, we only need to know the beginning and ending positions of the object, not all the information about how it got from one position to the other.

Self-Check

A ball hits the floor, bounces to a height of one meter, falls, and hits the floor again. Is the Δx between the two impacts equal to zero, one, or two meters?

Frames of reference

The example above shows that there are two arbitrary choices you have to make in order to define a position variable, x. You have to decide where to put $x=0$, and also which direction will be positive. This is referred to as choosing a *coordinate system* or choosing a *frame of reference*. (The two terms are nearly synonymous, but the first focuses more on the actual x variable, while the second is more of a general way of referring to one's point of view.) As long as you are consistent, any frame is equally valid. You just don't want to change coordinate systems in the middle of a calculation.

Have you ever been sitting in a train in a station when suddenly you notice that the station is moving backward? Most people would describe the situation by saying that you just failed to notice that the train was moving — it only seemed like the station was moving. But this shows that there is yet a third arbitrary choice that goes into choosing a coordinate system: valid frames of reference can differ from each other by moving relative to one another. It might seem strange that anyone would bother with a coordinate system that was moving relative to the earth, but for instance the frame of reference moving along with a train might be far more convenient for describing things happening inside the train.

Zero, because the "after" and "before" values of x are the same.

2.3 Graphs of Motion; Velocity.

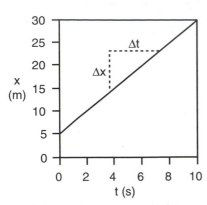

(a) Motion with constant velocity.

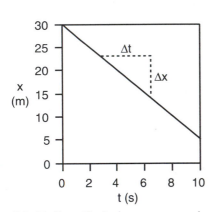

(b) Motion that decreases x is represented with negative values of Δx and v.

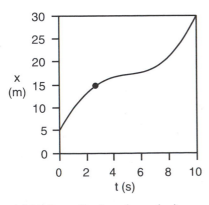

(c) Motion with changing velocity.

Motion with constant velocity

In example (a), an object is moving at constant speed in one direction. We can tell this because every two seconds, its position changes by five meters.

In algebra notation, we'd say that the graph of x vs. t shows the same change in position, $\Delta x = 5.0$ m, over each interval of $\Delta t = 2.0$ s. The object's velocity or speed is obtained by calculating $v = \Delta x / \Delta t = (5.0 \text{ m})/(2.0 \text{ s}) = 2.5$ m/s. In graphical terms, the velocity can be interpreted as the slope of the line. Since the graph is a straight line, it wouldn't have mattered if we'd taken a longer time interval and calculated $v = \Delta x / \Delta t = (10.0 \text{ m})/(4.0 \text{ s})$. The answer would still have been the same, 2.5 m/s.

Note that when we divide a number that has units of meters by another number that has units of seconds, we get units of meters per second, which can be written m/s. This is another case where we treat units as if they were algebra symbols, even though they're not.

In example (b), the object is moving in the opposite direction: as time progresses, its x coordinate decreases. Recalling the definition of the Δ notation as "after minus before," we find that Δt is still positive, but Δx must be negative. The slope of the line is therefore negative, and we say that the object has a negative velocity, $v = \Delta x / \Delta t = (-5.0 \text{ m})/(2.0 \text{ s}) = -2.5$ m/s. We've already seen that the plus and minus signs of Δx values have the interpretation of telling us which direction the object moved. Since Δt is always positive, dividing by Δt doesn't change the plus or minus sign, and the plus and minus signs of velocities are to be interpreted in the same way. In graphical terms, a positive slope characterizes a line that goes up as we go to the right, and a negative slope tells us that the line went down as we went to the right.

Motion with changing velocity

Now what about a graph like example (c)? This might be a graph of a car's motion as the driver cruises down the freeway, then slows down to look at a car crash by the side of the road, and then speeds up again, disappointed that there is nothing dramatic going on such as flames or babies trapped in their car seats. (Note that we are still talking about one-dimensional motion. Just because the graph is curvy doesn't mean that the car's path is curvy. The graph is not like a map, and the horizontal direction of the graph represents the passing of time, not distance.)

Example (c) is similar to example (a) in that the object moves a total of 25.0 m in a period of 10.0 s, but it is no longer true that it makes the same amount of progress every second. There is no way to characterize the entire graph by a certain velocity or slope, because the velocity is different at every moment. It would be incorrect to say that because the car covered 25.0 m in 10.0 s, its velocity was 2.5 m/s . It moved faster than that at the beginning and end, but slower in the middle. There may have been certain instants at which the car was indeed going 2.5 m/s, but the speedometer swept past that value without "sticking," just as it swung through various other values of speed. (I definitely want my next car to have a speedometer calibrated in m/s and showing both negative and positive values.)

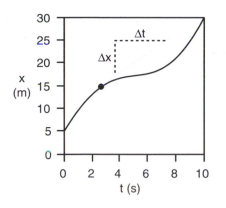

(d) The velocity at any given moment is defined as the slope of the tangent line through the relevant point on the graph.

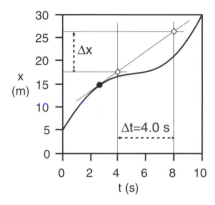

Example: finding the velocity at the point indicated with the dot.

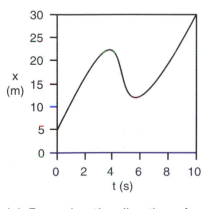

(e) Reversing the direction of motion.

We assume that our speedometer tells us what is happening to the speed of our car at every instant, but how can we define speed mathematically in a case like this? We can't just define it as the slope of the curvy graph, because a curve doesn't have a single well-defined slope as does a line. A mathematical definition that corresponded to the speedometer reading would have to be one that attached a different velocity value to a single point on the curve, i.e. a single instant in time, rather than to the entire graph. If we wish to define the speed at one instant such as the one marked with a dot, the best way to proceed is illustrated in (d), where we have drawn the line through that point called the tangent line, the line that "hugs the curve." We can then adopt the following definition of velocity:

---

definition of velocity

The velocity of an object at any given moment is the slope of the tangent line through the relevant point on its x-t graph.

One interpretation of this definition is that the velocity tells us how many meters the object would have traveled in one second, if it had continued moving at the same speed for at least one second. To some people the graphical nature of this definition seems "inaccurate" or "not mathematical." The equation $v=\Delta x/\Delta t$ by itself, however, is only valid if the velocity is constant, and so cannot serve as a general definition.

Example
Question: What is the velocity at the point shown with a dot on the graph?
Solution: First we draw the tangent line through that point. To find the slope of the tangent line, we need to pick two points on it. Theoretically, the slope should come out the same regardless of which two points we picked, but in practical terms we'll be able to measure more accurately if we pick two points fairly far apart, such as the two white diamonds. To save work, we pick points that are directly above labeled points on the t axis, so that Δt=4.0 s is easy to read off. One diamond lines up with $x \approx$17.5 m, the other with $x \approx$26.5 m, so Δx=9.0 m. The velocity is $\Delta x/\Delta t$=2.2 m/s.

Conventions about graphing

The placement of t on the horizontal axis and x on the upright axis may seem like an arbitrary convention, or may even have disturbed you, since your algebra teacher always told you that x goes on the horizontal axis and y goes on the upright axis. There is a reason for doing it this way, however. In example (e), we have an object that reverses its direction of motion twice. It can only be in one place at any given time, but there can be more than one time when it is at a given place. For instance, this object passed through x=17 m on three separate occasions, but there is no way it could have been in more than one place at t=5.0 s. Resurrecting some terminology you learned in your trigonometry course, we say that x is a function of t, but t is not a function of x. In situations such as this, there is a useful convention that the graph should be oriented so that any vertical line passes through the curve at only one point. Putting the x axis across the page and t upright would have violated this convention. To people who are used to interpreting graphs, a graph that violates this convention is as annoying as fingernails scratching on a chalkboard. We say that this is a graph of "x versus *t*." If the axes were the other way around, it would be a graph of "*t* versus *x*." I remember the "versus" terminology by visualizing the labels on the *x* and *t* axes and remembering that when you read, you go from left to right and from top to bottom.

Discussion questions

A. An ant walks forward, pauses, then runs quickly ahead. It then suddenly reverses direction and walks slowly back the way it came. Which graph could represent its motion?

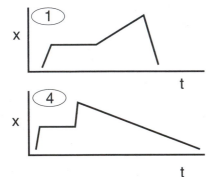

B. The figure shows a sequence of positions for two racing tractors. Compare the tractors' velocities as the race progresses. When do they have the same velocity?

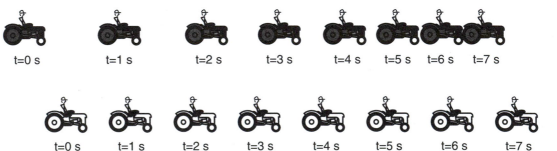

C. If an object had a straight-line motion graph with $\Delta x=0$ and $\Delta t \neq 0$, what would be true about its velocity? What would this look like on a graph? What about $\Delta t=0$ and $\Delta x \neq 0$?

D. If an object has a wavy motion graph like the one in example (e) on the previous page, which are the points at which the object reverses its direction? What is true about the object's velocity at these points?

E. Discuss anything unusual about the following three graphs.

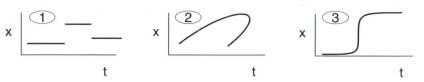

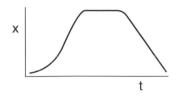

Discussion question G.

F. I have been using the term "velocity" and avoiding the more common English word "speed," because some introductory physics texts define them to mean different things. They use the word "speed," and the symbol "s" to mean the absolute value of the velocity, $s=|v|$. Although I have thrown in my lot with the minority of books that don't emphasize this distinction in technical vocabulary, there are clearly two different concepts here. Can you think of an example of a graph of x vs. t in which the object has constant speed, but not constant velocity?

G. In the graph on the left, describe how the object's velocity changes.

H. Two physicists duck out of a boring scientific conference to go get beer. On the way to the bar, they witness an accident in which a pedestrian is injured by a hit-and-run driver. A criminal trial results, and they must testify. In her testimony, Dr. Transverz Waive says, "The car was moving along pretty fast, I'd say the velocity was +40 mi/hr. They saw the old lady too late, and even though they slammed on the brakes they still hit her before they stopped. Then they made a U turn and headed off at a velocity of about -20 mi/hr, I'd say." Dr. Longitud N.L. Vibrasheun says, "He was really going too fast, maybe his velocity was -35 or -40 mi/hr. After he hit Mrs. Hapless, he turned around and left at a velocity of, oh, I'd guess maybe +20 or +25 mi/hr." Is their testimony contradictory? Explain.

2.4 The Principle of Inertia

Physical effects relate only to a change in velocity

Consider two statements that were at one time made with the utmost seriousness:

People like Galileo and Copernicus who say the earth is rotating must be crazy. We know the earth can't be moving. Why, if the earth was really turning once every day, then our whole city would have to be moving hundreds of leagues in an hour. That's impossible! Buildings would shake on their foundations. Gale-force winds would knock us over. Trees would fall down. The Mediterranean would come sweeping across the east coasts of Spain and Italy. And furthermore, what force would be making the world turn?

All this talk of passenger trains moving at forty miles an hour is sheer hogwash! At that speed, the air in a passenger compartment would all be forced against the back wall. People in the front of the car would suffocate, and people at the back would die because in such concentrated air, they wouldn't be able to expel a breath.

Some of the effects predicted in the first quote are clearly just based on a lack of experience with rapid motion that is smooth and free of vibration. But there is a deeper principle involved. In each case, the speaker is assuming that the mere fact of motion must have dramatic physical effects. More subtly, they also believe that a force is needed to keep an object in motion: the first person thinks a force would be needed to maintain the earth's rotation, and the second apparently thinks of the rear wall as pushing on the air to keep it moving.

Common modern knowledge and experience tell us that these people's predictions must have somehow been based on incorrect reasoning, but it is not immediately obvious where the fundamental flaw lies. It's one of those things a four-year-old could infuriate you by demanding a clear explanation of. One way of getting at the fundamental principle involved is to consider how the modern concept of the universe differs from the popular conception at the time of the Italian Renaissance. To us, the word "earth" implies a planet, one of the nine planets of our solar system, a small ball of rock and dirt that is of no significance to anyone in the universe except for members of our species, who happen to live on it. To Galileo's contemporaries, however, the earth was the biggest, most solid, most important thing in all of creation, not to be compared with the wandering lights in the sky known as planets. To us, the earth is just another object, and when we talk loosely about "how fast" an object such as a car "is going," we really mean the car-object's velocity relative to the earth-object.

Motion is relative

According to our modern world-view, it really isn't that reasonable to expect that a special force should be required to make the air in the train have a certain velocity relative to our planet. After all, the "moving" air in the "moving" train might just happen to have zero velocity relative to some other planet we don't even know about. Aristotle claimed that things "naturally" wanted to be at rest, lying on the surface of the earth. But experiment after experiment has shown that there is really nothing so

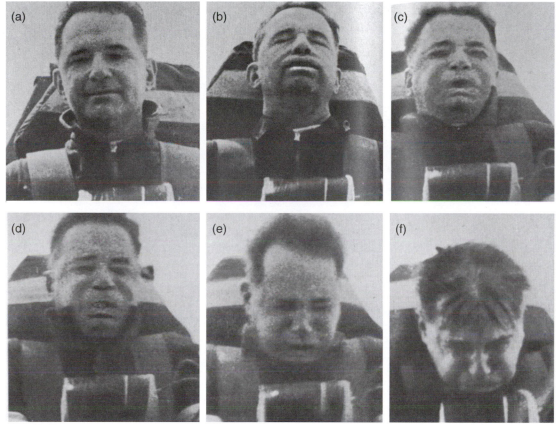

This Air Force doctor volunteered to ride a rocket sled as a medical experiment. The obvious effects on his head and face are not because of the sled's speed but because of its rapid changes in speed: increasing in (b) and (c), and decreasing in (e) and (f). In (d) his speed is greatest, but because his speed is not increasing or decreasing very much at this moment, there is little effect on him.

special about being at rest relative to the earth. For instance, if a mattress falls out of the back of a truck on the freeway, the reason it rapidly comes to rest with respect to the planet is simply because of friction forces exerted by the asphalt, which happens to be attached to the planet.

Galileo's insights are summarized as follows:

The Principle of Inertia

No force is required to maintain motion with constant velocity in a straight line, and absolute motion does not cause any observable physical effects.

There are many examples of situations that seem to disprove the principle of inertia, but these all result from forgetting that friction is a force. For instance, it seems that a force is needed to keep a sailboat in motion. If the wind stops, the sailboat stops too. But the wind's force is not the only force on the boat; there is also a frictional force from the water. If the sailboat is cruising and the wind suddenly disappears, the backward frictional force still exists, and since it is no longer being counteracted by the wind's forward force, the boat stops. To disprove the principle of inertia, we would have to find an example where a moving object slowed down even though no forces whatsoever were acting on it.

Self-Check

What is incorrect about the following supposed counterexamples to the principle of inertia?

(1) When astronauts blast off in a rocket, their huge velocity does cause a physical effect on their bodies — they get pressed back into their seats, the flesh on their faces gets distorted, and they have a hard time lifting their arms.

(2) When you're driving in a convertible with the top down, the wind in your face is an observable physical effect of your absolute motion.

Discussion questions

A. A passenger on a cruise ship finds, while the ship is docked, that he can leap off of the upper deck and just barely make it into the pool on the lower deck. If the ship leaves dock and is cruising rapidly, will this adrenaline junkie still be able to make it?

B. You are a passenger in the open basket hanging under a helium balloon. The balloon is being carried along by the wind at a constant velocity. If you are holding a flag in your hand, will the flag wave? If so, which way? [Based on a question from PSSC Physics.]

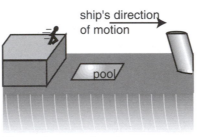

Discussion question A.

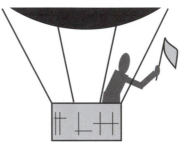

Discussion question B.

C. Aristotle stated that all objects naturally wanted to come to rest, with the unspoken implication that "rest" would be interpreted relative to the surface of the earth. Suppose we go back in time and transport Aristotle to the moon. Aristotle knew, as we do, that the moon circles the earth; he said it didn't fall down because, like everything else in the heavens, it was made out of some special substance whose "natural" behavior was to go in circles around the earth. We land, put him in a space suit, and kick him out the door. What would he expect his fate to be in this situation? If intelligent creatures inhabited the moon, and one of them independently came up with the equivalent of Aristotelian physics, what would they think about objects coming to rest?

D. The bottle is sitting on a level table in a train's dining car, but the surface of the beer is tilted. What can you infer about the motion of the train?

Discussion question D.

---

(1) The effect only occurs during blastoff, when their velocity is changing. Once the rocket engines stop firing, their velocity stops changing, and they no longer feel any effect. (2) It is only an observable effect of your motion relative to the air.

2.5 Addition of Velocities

Addition of velocities to describe relative motion

Since absolute motion cannot be unambiguously measured, the only way to describe motion unambiguously is to describe the motion of one object relative to another. Symbolically, we can write v_{PQ} for the velocity of object P relative to object Q.

Velocities measured with respect to different reference points can be compared by addition. In the figure below, the ball's velocity relative to the couch equals the ball's velocity relative to the truck plus the truck's velocity relative to the couch:

$$v_{BC} = v_{BT} + v_{TC}$$
$$= 5 \text{ cm/s} + 10 \text{ cm/s}$$

The same equation can be used for any combination of three objects, just by substituting the relevant subscripts for B, T, and C. Just remember to write the equation so that the velocities being added have the same subscript twice in a row. In this example, if you read off the subscripts going from left to right, you get BC...=...BTTC. The fact that the two "inside" subscripts on the right are the same means that the equation has been set up correctly. Notice how subscripts on the left look just like the subscripts on the right, but with the two T's eliminated.

These two highly competent physicists disagree on absolute velocities, but they would agree on relative velocities. Purple Dino considers the couch to be at rest, while Green Dino thinks of the truck as being at rest. They agree, however, that the truck's velocity relative to the couch is $v_{TC}=10$ cm/s, the ball's velocity relative to the truck is $v_{BT}=5$ cm/s, and the ball's velocity relative to the couch is $v_{BC}=v_{BT}+v_{TC}=15$ cm/s.

Negative velocities in relative motion

My discussion of how to interpret positive and negative signs of velocity may have left you wondering why we should bother. Why not just make velocity positive by definition? The original reason why negative numbers were invented was that bookkeepers decided it would be convenient to use the negative number concept for payments to distinguish them from receipts. It was just plain easier than writing receipts in black and payments in red ink. After adding up your month's positive receipts and negative payments, you either got a positive number, indicating profit, or a negative number, showing a loss. You could then show the that total with a high-tech "+" or "−" sign, instead of looking around for the appropriate bottle of ink.

Nowadays we use positive and negative numbers for all kinds of things, but in every case the point is that it makes sense to add and subtract those things according to the rules you learned in grade school, such as "minus a minus makes a plus, why this is true we need not discuss." Adding velocities has the significance of comparing relative motion, and with this interpretation negative and positive velocities can used within a consistent framework. For example, the truck's velocity relative to the couch equals the truck's velocity relative to the ball plus the ball's velocity relative to the couch:

$$\begin{aligned} v_{TC} &= v_{TB} + v_{BC} \\ &= -5 \text{ cm/s} + 15 \text{ cm/s} \\ &= 10 \text{ cm/s} \end{aligned}$$

If we didn't have the technology of negative numbers, we would have had to remember a complicated set of rules for adding velocities: (1) if the two objects are both moving forward, you add, (2) if one is moving forward and one is moving backward, you subtract, but (3) if they're both moving backward, you add. What a pain that would have been.

Discussion questions

A. Interpret the general rule $v_{AB} = -v_{BA}$ in words.

B. Wa-Chuen slips away from her father at the mall and walks up the down escalator, so that she stays in one place. Write this in terms of symbols.

2.6   Graphs of Velocity Versus Time

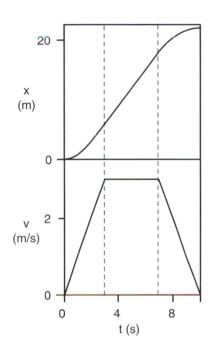

Since changes in velocity play such a prominent role in physics, we need a better way to look at changes in velocity than by laboriously drawing tangent lines on *x*-versus-*t* graphs. A good method is to draw a graph of velocity versus time. The examples on the left show the *x*-*t* and *v*-*t* graphs that might be produced by a car starting from a traffic light, speeding up, cruising for a while at constant speed, and finally slowing down for a stop sign. If you have an air freshener hanging from your rear-view mirror, then you will see an effect on the air freshener during the beginning and ending periods when the velocity is changing, but it will not be tilted during the period of constant velocity represented by the flat plateau in the middle of the *v*-*t* graph.

Students often mix up the things being represented on these two types of graphs. For instance, many students looking at the top graph say that the car is speeding up the whole time, since "the graph is becoming greater." What is getting greater throughout the graph is *x*, not *v*.

Similarly, many students would look at the bottom graph and think it showed the car backing up, because "it's going backwards at the end." But what is decreasing at the end is *v*, not *x*. Having both the *x*-*t* and *v*-*t* graphs in front of you like this is often convenient, because one graph may be easier to interpret than the other for a particular purpose. Stacking them like this means that corresponding points on the two graphs' time axes are lined up with each other vertically. However, one thing that is a little counterintuitive about the arrangement is that in a situation like this one involving a car, one is tempted to visualize the landscape stretching along the horizontal axis of one of the graphs. The horizontal axes, however, represent time, not position. The correct way to visualize the landscape is by mentally rotating the horizon 90 degrees counterclockwise and imagining it stretching along the upright axis of the *x*-*t* graph, which is the only axis that represents different positions in space.

2.7 ∫   Applications of Calculus

The integral symbol, ∫, in the heading for this section indicates that it is meant to be read by students in calculus-based physics. Students in an algebra-based physics course should skip these sections. The calculus-related sections in this book are meant to be usable by students who are taking calculus concurrently, so at this early point in the physics course I do not assume you know any calculus yet. This section is therefore not much more than a quick preview of calculus, to help you relate what you're learning in the two courses.

Newton was the first person to figure out the tangent-line definition of velocity for cases where the *x*-*t* graph is nonlinear. Before Newton, nobody had conceptualized the description of motion in terms of *x*-*t* and *v*-*t* graphs. In addition to the graphical techniques discussed in this chapter, Newton also invented a set of symbolic techniques called calculus. If you have an equation for *x* in terms of *t*, calculus allows you, for instance, to find an equation for *v* in terms of *t*. In calculus terms, we say that the function *v*(*t*) is the derivative of the function $x(t)$. In other words, the derivative of a function is a new function that tells how rapidly the original function was changing. We now use neither Newton's name for his technique (he called it "the method of fluxions") nor his notation. The more commonly used notation is due to Newton's German contemporary Leibnitz, whom the English accused of plagiarizing the calculus from Newton. In the Leibnitz notation, we write

$$v = \frac{dx}{dt}$$

to indicate that the function $v(t)$ equals the slope of the tangent line of the graph of $x(t)$ at every time t. The Leibnitz notation is meant to evoke the delta notation, but with a very small time interval. Because the dx and dt are thought of as very small Δx's and Δt's, i.e. very small differences, the part of calculus that has to do with derivatives is called differential calculus.

Differential calculus consists of three things:

- The concept and definition of the derivative, which is covered in this book, but which will be discussed more formally in your math course.

- The Leibnitz notation described above, which you'll need to get more comfortable with in your math course.

- A set of rules for that allows you to find an equation for the derivative of a given function. For instance, if you happened to have a situation where the position of an object was given by the equation $x=2t^7$, you would be able to use those rules to find $dx/dt=14t^6$. This bag of tricks is covered in your math course.

Summary

Selected Vocabulary

center of mass .... the balance point of an object

velocity .... the rate of change of position; the slope of the tangent line on an *x-t* graph.

Notation

x .... a point in space

t .... a point in time, a clock reading

Δ .... "change in;" the value of a variable afterwards minus its value before

Δx .... a distance, or more precisely a change in x, which may be less than the distance traveled; its plus or minus sign indicates direction

Δt .... a duration of time

v .... velocity

v_{AB} .... the velocity of object A relative to object B

Standard Terminology Avoided in This Book

displacement .... a name for the symbol Δx.

speed .... the absolute value of the velocity, i.e. the velocity stripped of any information about its direction

Summary

An object's center of mass is the point at which it can be balanced. For the time being, we are studying the mathematical description only of the motion of an object's center of mass in cases restricted to one dimension. The motion of an object's center of mass is usually far simpler than the motion of any of its other parts.

It is important to distinguish location, x, from distance, Δx, and clock reading, t, from time interval Δt. When an object's *x-t* graph is linear, we define its velocity as the slope of the line, $\Delta x/\Delta t$. When the graph is curved, we generalize the definition so that the velocity is the slope of the tangent line at a given point on the graph.

Galileo's principle of inertia states that no force is required to maintain motion with constant velocity in a straight line, and absolute motion does not cause any observable physical effects. Things typically tend to reduce their velocity relative to the surface of our planet only because they are physically rubbing against the planet (or something attached to the planet), not because there is anything special about being at rest with respect to the earth's surface. When it seems, for instance, that a force is required to keep a book sliding across a table, in fact the force is only serving to cancel the contrary force of friction.

Absolute motion is not a well-defined concept, and if two observers are not at rest relative to one another they will disagree about the absolute velocities of objects. They will, however, agree about relative velocities. If object A is in motion relative to object B, and B is in motion relative to C, then A's velocity relative to C is given by $v_{AC}=v_{AB}+v_{BC}$. Positive and negative signs are used to indicate the direction of an object's motion.

Homework Problems

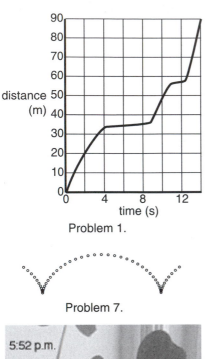

Problem 1.

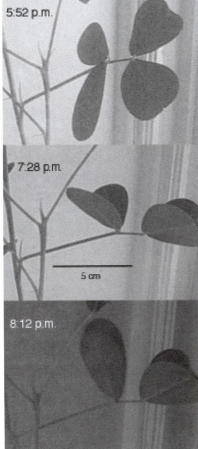

Problem 7.

Problem 8.

1 ✓. The graph shows the motion of a car stuck in stop-and-go freeway traffic. (a) If you only knew how far the car had gone during this entire time period, what would you think its velocity was? (b) What is the car's maximum velocity?

2. (a) Let θ be the latitude of a point on the Earth's surface. Derive an algebra equation for the distance, L, traveled by that point during one rotation of the Earth about its axis, i.e. over one day, expressed in terms of L, θ, and R, the radius of the earth. Check: Your equation should give $L=0$ for the North Pole.

(b✓) At what speed is Fullerton, at latitude θ=34°, moving with the rotation of the Earth about its axis? Give your answer in units of mi/h. [See the table in the back of the book for the relevant data.]

3★✓. A person is parachute jumping. During the time between when she leaps out of the plane and when she opens her chute, her altitude is given by the equation

$$y=(10000 \text{ m}) - (50 \text{ m/s})[t+(5.0 \text{ s})e^{-t/5.0 \text{ s}}] \quad .$$

Find her velocity at $t=7.0$ s. (This can be done on a calculator, without knowing calculus.) Because of air resistance, her velocity does not increase at a steady rate as it would for an object falling in vacuum.

4 S. A light-year is a unit of distance used in astronomy, and defined as the distance light travels in one year. The speed of light is 3.0×10^8 m/s. Find how many meters there are in one light-year, expressing your answer in scientific notation.

5 S. You're standing in a freight train, and have no way to see out. If you have to lean to stay on your feet, what, if anything, does that tell you about the train's velocity? Its acceleration? Explain.

6 ∫. A honeybee's position as a function of time is given by $x=10t-t^3$, where t is in seconds and x in meters. What is its velocity at $t=3.0$ s?

7 S. The figure shows the motion of a point on the rim of a rolling wheel. (The shape is called a cycloid.) Suppose bug A is riding on the rim of the wheel on a bicycle that is rolling, while bug B is on the spinning wheel of a bike that is sitting upside down on the floor. Bug A is moving along a cycloid, while bug B is moving in a circle. Both wheels are doing the same number of revolutions per minute. Which bug has a harder time holding on, or do they find it equally difficult?

8 ✓. Peanut plants fold up their leaves at night. Estimate the top speed of the tip of one of the leaves shown in the figure, expressing your result in scientific notation in SI units.

S   A solution is given in the back of the book.   ★ A difficult problem.
✓ A computerized answer check is available.   ∫ A problem that requires calculus.

9. (a) Translate the following information into symbols, using the notation with two subscripts introduced in section 2.5. Eowyn is riding on her horse at a velocity of 11 m/s. She twists around in her saddle and fires an arrow backward. Her bow fires arrows at 25 m/s. (b) Find the speed of the arrow relative to the ground.

10 S. Our full discussion of two- and three-dimensional motion is postponed until the second half of the book, but here is a chance to use a little mathematical creativity in anticipation of that generalization. Suppose a ship is sailing east at a certain speed v, and a passenger is walking across the deck at the same speed v, so that his track across the deck is perpendicular to the ship's center-line. What is his speed relative to the water, and in what direction is he moving relative to the water?

11∫. Freddi Fish$^{(TM)}$ has a position as a function of time given by $x = a/(b+t^2)$. Find her maximum speed.

3 Acceleration and Free Fall

3.1 The Motion of Falling Objects

Galileo dropped a cannonball and a musketball simultaneously from a tower, and observed that they hit the ground at nearly the same time. This contradicted Aristotle's long-accepted idea that heavier objects fell faster.

The motion of falling objects is the simplest and most common example of motion with changing velocity. The early pioneers of physics had a correct intuition that the way things drop was a message directly from Nature herself about how the universe worked. Other examples seem less likely to have deep significance. A walking person who speeds up is making a conscious choice. If one stretch of a river flows more rapidly than another, it may be only because the channel is narrower there, which is just an accident of the local geography. But there is something impressively consistent, universal, and inexorable about the way things fall.

Stand up now and simultaneously drop a coin and a bit of paper side by side. The paper takes much longer to hit the ground. That's why Aristotle wrote that heavy objects fell more rapidly. Europeans believed him for two thousand years.

Now repeat the experiment, but make it into a race between the coin and your shoe. My own shoe is about 50 times heavier than the nickel I had handy, but it looks to me like they hit the ground at exactly the same moment. So much for Aristotle! Galileo, who had a flair for the theatrical, did the experiment by dropping a bullet and a heavy cannonball from a tall tower. Aristotle's observations had been incomplete, his interpretation a vast oversimplification.

It is inconceivable that Galileo was the first person to observe a discrepancy with Aristotle's predictions. Galileo was the one who changed the course of history because he was able to assemble the observations into a coherent pattern, and also because he carried out systematic quantitative (numerical) measurements rather than just describing things qualitatively.

Why is it that some objects, like the coin and the shoe, have similar motion, but others, like a feather or a bit of paper, are different? Galileo

Galileo and the Church

Galileo's contradiction of Aristotle had serious consequences. He was interrogated by the Church authorities and convicted of teaching that the earth went around the sun as a matter of fact and not, as he had promised previously, as a mere mathematical hypothesis. He was placed under permanent house arrest, and forbidden to write about or teach his theories. Immediately after being forced to recant his claim that the earth revolved around the sun, the old man is said to have muttered defiantly "and yet it does move."

The story is dramatic, but there are some omissions in the commonly taught heroic version. There was a rumor that the Simplicio character represented the Pope. Also, some of the ideas Galileo advocated had controversial religious overtones. He believed in the existence of atoms, and atomism was thought by some people to contradict the Church's doctrine of transubstantiation, which said that in the Catholic mass, the blessing of the bread and wine literally transformed them into the flesh and blood of Christ. His support for a cosmology in which the earth circled the sun was also disreputable because one of its supporters, Giordano Bruno, had also proposed a bizarre synthesis of Christianity with the ancient Egyptian religion.

speculated that in addition to the force that always pulls objects down, there was an upward force exerted by the air. Anyone can speculate, but Galileo went beyond speculation and came up with two clever experiments to probe the issue. First, he experimented with objects falling in water, which probed the same issues but made the motion slow enough that he could take time measurements with a primitive pendulum clock. With this technique, he established the following facts:

- All heavy, streamlined objects (for example a steel rod dropped point-down) reach the bottom of the tank in about the same amount of time, only slightly longer than the time they would take to fall the same distance in air.
- Objects that are lighter or less streamlined take a longer time to reach the bottom.

This supported his hypothesis about two contrary forces. He imagined an idealized situation in which the falling object did not have to push its way through any substance at all. Falling in air would be more like this ideal case than falling in water, but even a thin, sparse medium like air would be sufficient to cause obvious effects on feathers and other light objects that were not streamlined. Today, we have vacuum pumps that allow us to suck nearly all the air out of a chamber, and if we drop a feather and a rock side by side in a vacuum, the feather does not lag behind the rock at all.

How the speed of a falling object increases with time

Galileo's second stroke of genius was to find a way to make quantitative measurements of how the speed of a falling object increased as it went along. Again it was problematic to make sufficiently accurate time measurements with primitive clocks, and again he found a tricky way to slow things down while preserving the essential physical phenomena: he let a ball roll down a slope instead of dropping it vertically. The steeper the incline, the more rapidly the ball would gain speed. Without a modern video camera, Galileo had invented a way to make a slow-motion version of falling.

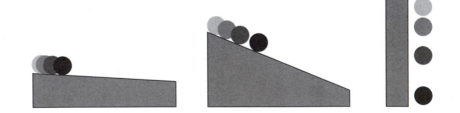

Velocity increases more gradually on the gentle slope, but the motion is otherwise the same as the motion of a falling object.

Although Galileo's clocks were only good enough to do accurate experiments at the smaller angles, he was confident after making a systematic study at a variety of small angles that his basic conclusions were generally valid. Stated in modern language, what he found was that the velocity-versus-time graph was a line. In the language of algebra, we know that a line has an equation of the form $y=ax+b$, but our variables are v and t, so it would be $v=at+b$. (The constant b can be interpreted simply as the initial velocity of the object, i.e. its velocity at the time when we started our clock, which we conventionally write as v_o.)

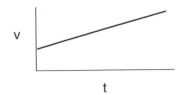

The v-t graph of a falling object is a line.

Self-Check

An object is rolling down an incline. After it has been rolling for a short time, it is found to travel 13 cm during a certain one-second interval. During the second after that, if goes 16 cm. How many cm will it travel in the second after that?

(a) Galileo's experiments show that all falling objects have the same motion if air resistance is negligible.

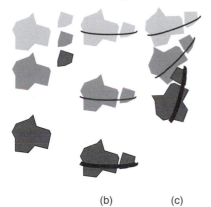

(b)   (c)

Aristotle said that heavier objects fell faster than lighter ones. If two rocks are tied together, that makes an extra-heavy rock, (b), which should fall faster. But Aristotle's theory would also predict that the light rock would hold back the heavy rock, resulting in a slower fall, (c).

A contradiction in Aristotle's reasoning

Galileo's inclined-plane experiment disproved the long-accepted claim by Aristotle that a falling object had a definite "natural falling speed" proportional to its weight. Galileo had found that the speed just kept on increasing, and weight was irrelevant as long as air friction was negligible. Not only did Galileo prove experimentally that Aristotle had been wrong, but he also pointed out a logical contradiction in Aristotle's own reasoning. Simplicio, the stupid character, mouths the accepted Aristotelian wisdom:

SIMPLICIO: There can be no doubt but that a particular body ... has a fixed velocity which is determined by nature...

SALVIATI: If then we take two bodies whose natural speeds are different, it is clear that, [according to Aristotle], on uniting the two, the more rapid one will be partly held back by the slower, and the slower will be somewhat hastened by the swifter. Do you not agree with me in this opinion?

SIMPLICIO: You are unquestionably right.

SALVIATI: But if this is true, and if a large stone moves with a speed of, say, eight [unspecified units] while a smaller moves with a speed of four, then when they are united, the system will move with a speed less than eight; but the two stones when tied together make a stone larger than that which before moved with a speed of eight. Hence the heavier body moves with less speed than the lighter; an effect which is contrary to your supposition. Thus you see how, from your assumption that the heavier body moves more rapidly than the lighter one, I infer that the heavier body moves more slowly.

[tr. Crew and De Salvio]

What is gravity?

The physicist Richard Feynman liked to tell a story about how when he was a little kid, he asked his father, "Why do things fall?" As an adult, he praised his father for answering, "Nobody knows why things fall. It's a deep mystery, and the smartest people in the world don't know the basic reason for it." Contrast that with the average person's off-the-cuff answer, "Oh, it's because of gravity." Feynman liked his father's answer, because his father realized that simply giving a name to something didn't mean that you understood it. The radical thing about Galileo's and Newton's approach to science was that they concentrated first on describing mathematically what really did happen, rather than spending a lot of time on untestable speculation such as Aristotle's statement that "Things fall because they are trying to reach their natural place in contact with the earth." That doesn't mean that science can never answer the "why" questions. Over the next month or two as you delve deeper into physics, you will learn that there are more fundamental reasons why all falling objects have v-t graphs with the same slope, regardless of their mass. Nevertheless, the methods of science always impose limits on how deep our explanation can go.

Its speed increases at a steady rate, so in the next second it will travel 19 cm.

3.2 Acceleration

Definition of acceleration for linear v-t graphs

Galileo's experiment with dropping heavy and light objects from a tower showed that all falling objects have the same motion, and his inclined-plane experiments showed that the motion was described by $v=ax+v_o$. The initial velocity v_o depends on whether you drop the object from rest or throw it down, but even if you throw it down, you cannot change the slope, a, of the v-t graph.

Since these experiments show that all falling objects have linear v-t graphs with the same slope, the slope of such a graph is apparently an important and useful quantity. We use the word *acceleration*, and the symbol a, for the slope of such a graph. In symbols, $a=\Delta v/\Delta t$. The acceleration can be interpreted as the amount of speed gained in every second, and it has units of velocity divided by time, i.e. "meters per second per second," or m/s/s. Continuing to treat units as if they were algebra symbols, we simplify "m/s/s" to read "m/s^2." Acceleration can be a useful quantity for describing other types of motion besides falling, and the word and the symbol "a" can be used in a more general context. We reserve the more specialized symbol "g" for the acceleration of falling objects, which on the surface of our planet equals 9.8 m/s^2. Often when doing approximate calculations or merely illustrative numerical examples it is good enough to use $g=10$ m/s^2, which is off by only 2%.

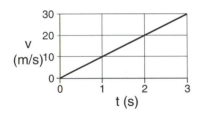

Example
Question: A despondent physics student jumps off a bridge, and falls for three seconds before hitting the water. How fast is he going when he hits the water?
Solution: Approximating g as 10 m/s^2, he will gain 10 m/s of speed each second. After one second, his velocity is 10 m/s, after two seconds it is 20 m/s, and on impact, after falling for three seconds, he is moving at 30 m/s.

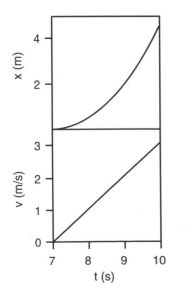

Example: extracting acceleration from a graph
Question: The x-t and v-t graphs show the motion of a car starting from a stop sign. What is the car's acceleration?
Solution: Acceleration is defined as the slope of the v-t graph. The graph rises by 3 m/s during a time interval of 3 s, so the acceleration is (3 m/s)/(3 s)=1 m/s^2.
Incorrect solution #1: The final velocity is 3 m/s, and acceleration is velocity divided by time, so the acceleration is (3 m/s)/(10 s)=0.3 m/s^2.
✗ The solution is incorrect because you can't find the slope of a graph from one point. This person was just using the point at the right end of the v-t graph to try to find the slope of the curve.
Incorrect solution #2: Velocity is distance divided by time so v=(4.5 m)/(3 s)=1.5 m/s. Acceleration is velocity divided by time, so a=(1.5 m/s)/(3 s)=0.5 m/s^2.
✗ The solution is incorrect because velocity is the slope of the tangent line. In a case like this where the velocity is changing, you can't just pick two points on the x-t graph and use them to find the velocity.

Example: converting g to different units
Question: What is g in units of cm/s²?
Solution: The answer is going to be how many cm/s of speed a falling object gains in one second. If it gains 9.8 m/s in one second, then it gains 980 cm/s in one second, so g=980 cm/s². Alternatively, we can use the method of fractions that equal one:

$$\frac{9.8 \ \cancel{m}}{s^2} \times \frac{100 \text{ cm}}{1 \ \cancel{m}} = \frac{980 \text{ cm}}{s^2}$$

Question: What is g in units of miles/hour²?
Solution:

$$\frac{9.8 \text{ m}}{s^2} \times \frac{1 \text{ mile}}{1600 \text{ m}} \times \left(\frac{3600 \text{ s}}{1 \text{ hour}}\right)^2 = 7.9 \times 10^4 \text{ mile / hour}^2$$

This large number can be interpreted as the speed, in miles per hour, you would gain by falling for one hour. Note that we had to square the conversion factor of 3600 s/hour in order to cancel out the units of seconds squared in the denominator.

Question: What is g in units of miles/hour/s?
Solution:

$$\frac{9.8 \text{ m}}{s^2} \times \frac{1 \text{ mile}}{1600 \text{ m}} \times \frac{3600 \text{ s}}{1 \text{ hour}} = 22 \text{ mile/hour/s}$$

This is a figure that Americans will have an intuitive feel for. If your car has a forward acceleration equal to the acceleration of a falling object, then you will gain 22 miles per hour of speed every second. However, using mixed time units of hours and seconds like this is usually inconvenient for problem-solving. It would be like using units of foot-inches for area instead of ft² or in².

The acceleration of gravity is different in different locations.

Everyone knows that gravity is weaker on the moon, but actually it is not even the same everywhere on Earth, as shown by the sampling of numerical data in the following table.

location	latitude	elevation (m)	g (m/s²)
north pole	90° N	0	9.8322
Reykjavik, Iceland	64° N	0	9.8225
Fullerton, California	34° N	0	9.7957
Guayaquil, Ecuador	2° S	0	9.7806
Mt. Cotopaxi, Ecuador	1° S	5896	9.7624
Mt. Everest	28° N	8848	9.7643

The main variables that relate to the value of *g* on Earth are latitude and elevation. Although you have not yet learned how *g* would be calculated based on any deeper theory of gravity, it is not too hard to guess why *g* depends on elevation. Gravity is an attraction between things that have mass, and the attraction gets weaker with increasing distance. As you ascend from the seaport of Guayaquil to the nearby top of Mt. Cotopaxi, you are distancing yourself from the mass of the planet. The dependence on latitude occurs because we are measuring the acceleration of gravity relative to the earth's surface, but the earth's rotation causes the earth's surface to fall out from under you. (We will discuss both gravity and rotation in more detail later in the course.)

Much more spectacular differences in the strength of gravity can be observed away from the Earth's surface:

location	g (m/s^2)
asteroid Vesta (surface)	0.3
Earth's moon (surface)	1.6
Mars (surface)	3.7
Earth (surface)	9.8
Jupiter (cloud-tops)	26
Sun (visible surface)	270
typical neutron star (surface)	10^{12}
black hole (center)	infinite according to some theories, on the order of 10^{52} according to others

A typical neutron star is not so different in size from a large asteroid, but is orders of magnitude more massive, so the mass of a body definitely correlates with the g it creates. On the other hand, a neutron star has about the same mass as our Sun, so why is its g billions of times greater? If you had the misfortune of being on the surface of a neutron star, you'd be within a few thousand miles of all its mass, whereas on the surface of the Sun, you'd still be millions of miles from most if its mass.

This false-color map shows variations in the strength of the earth's gravity. Purple areas have the strongest gravity, yellow the weakest. The overall trend toward weaker gravity at the equator and stronger gravity at the poles has been artificially removed to allow the weaker local variations to show up. The map covers only the oceans because of the technique used to make it: satellites look for bulges and depressions in the surface of the ocean. A very slight bulge will occur over an undersea mountain, for instance, because the mountain's gravitational attraction pulls water toward it. The US government originally began collecting data like these for military use, to correct for the deviations in the paths of missiles. The data have recently been released for scientific and commercial use (e.g. searching for sites for off-shore oil wells).

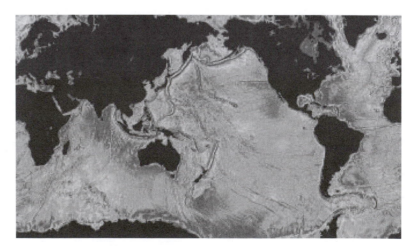

Discussion questions

A. What is wrong with the following definitions of g?

(a) "g is gravity."

(b) "g is the speed of a falling object."

(c) "g is how hard gravity pulls on things."

B. When advertisers specify how much acceleration a car is capable of, they do not give an acceleration as defined in physics. Instead, they usually specify how many seconds are required for the car to go from rest to 60 miles/hour. Suppose we use the notation "a" for the acceleration as defined in physics, and "$a_{\text{car ad}}$" for the quantity used in advertisements for cars. In the US's non-metric system of units, what would be the units of a and $a_{\text{car ad}}$? How would the use and interpretation of large and small, positive and negative values be different for a as opposed to $a_{\text{car ad}}$?

C. Two people stand on the edge of a cliff. As they lean over the edge, one person throws a rock down, while the other throws one straight up with an exactly opposite initial velocity. Compare the speeds of the rocks on impact at the bottom of the cliff.

3.3 Positive and Negative Acceleration

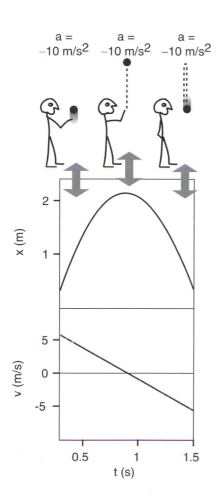

Gravity always pulls down, but that does not mean it always speeds things up. If you throw a ball straight up, gravity will first slow it down to $v=0$ and then begin increasing its speed. When I took physics in high school, I got the impression that positive signs of acceleration indicated speeding up, while negative accelerations represented slowing down, i.e. deceleration. Such a definition would be inconvenient, however, because we would then have to say that the same downward tug of gravity could produce either a positive or a negative acceleration. As we will see in the following example, such a definition also would not be the same as the slope of the v-t graph

Let's study the example of the rising and falling ball. In the example of the person falling from a bridge, I assumed positive velocity values without calling attention to it, which meant I was assuming a coordinate system whose x axis pointed down. In this example, where the ball is reversing direction, it is not possible to avoid negative velocities by a tricky choice of axis, so let's make the more natural choice of an axis pointing up. The ball's velocity will initially be a positive number, because it is heading up, in the same direction as the x axis, but on the way back down, it will be a negative number. As shown in the figure, the v-t graph does not do anything special at the top of the ball's flight, where v equals 0. Its slope is always negative. In the left half of the graph, the negative slope indicates a positive velocity that is getting closer to zero. On the right side, the negative slope is due to a negative velocity that is getting farther from zero, so we say that the ball is speeding up, but its velocity is decreasing!

To summarize, what makes the most sense is to stick with the original definition of acceleration as the slope of the v-t graph, $\Delta v/\Delta t$. By this definition, it just isn't necessarily true that things speeding up have positive acceleration while things slowing down have negative acceleration. The word "deceleration" is not used much by physicists, and the word "acceleration" is used unblushingly to refer to slowing down as well as speeding up: "There was a red light, and we accelerated to a stop."

Example

Question: In the example above, what happens if you calculate the acceleration between t=1.0 and 1.5 s?

Answer: Reading from the graph, it looks like the velocity is about -1 m/s at t=1.0 s, and around -6 m/s at t=1.5 s. The acceleration, figured between these two points, is

$$a = \frac{\Delta v}{\Delta t} = \frac{(-6 \text{ m/s}) - (-1 \text{ m/s})}{(1.5 \text{ s}) - (1.0 \text{ s})} = -10 \text{ m/s}^2 \quad .$$

Even though the ball is speeding up, it has a negative acceleration.

Another way of convincing you that this way of handling the plus and minus signs makes sense is to think of a device that measures acceleration. After all, physics is supposed to use operational definitions, ones that relate to the results you get with actual measuring devices. Consider an air freshener hanging from the rear-view mirror of your car. When you speed up, the air freshener swings backward. Suppose we define this as a positive reading. When you slow down, the air freshener swings forward, so we'll call this a negative reading on our accelerometer. But what if you put the car in reverse and start speeding up backwards? Even though you're speeding up, the accelerometer responds in the same way as it did when you were going forward and slowing down. There are four possible cases:

motion of car	accelerometer swings	slope of v-t graph	direction of force acting on car
forward, speeding up	backward	+	forward
forward, slowing down	forward	-	backward
backward, speeding up	forward	-	backward
backward, slowing down	backward	+	forward

Note the consistency of the three right-hand columns — nature is trying to tell us that this is the right system of classification, not the left-hand column.

Because the positive and negative signs of acceleration depend on the choice of a coordinate system, the acceleration of an object under the influence of gravity can be either positive or negative. Rather than having to write things like "g=9.8 m/s^2 or -9.8 m/s^2" every time we want to discuss g's numerical value, we simply define g as the absolute value of the acceleration of objects moving under the influence of gravity. We consistently let g=9.8 m/s^2, but we may have either a=g or a=-g, depending on our choice of a coordinate system.

Example

Question: A person kicks a ball, which rolls up a sloping street, comes to a halt, and rolls back down again. The ball has constant acceleration. The ball is initially moving at a velocity of 4.0 m/s, and after 10.0 s it has returned to where it started. At the end, it has sped back up to the same speed it had initially, but in the opposite direction. What was its acceleration?

Solution: By giving a positive number for the initial velocity, the statement of the question implies a coordinate axis that points up the slope of the hill. The "same" speed in the opposite direction should therefore be represented by a negative number, -4.0 m/s. The acceleration is a=$\Delta v/\Delta t$=(v_{after}-v_{before})/10.0 s=[(-4.0 m/s)-(4.0 m/s)]/10.0 s=-0.80 m/s^2. The acceleration was no different during the upward part of the roll than on the downward part of the roll.

Incorrect solution: Acceleration is $\Delta v/\Delta t$, and at the end it's not moving any faster or slower than when it started, so Δv=0 and a=0.

✗ The velocity does change, from a positive number to a negative number.

Discussion questions

A. A child repeatedly jumps up and down on a trampoline. Discuss the sign and magnitude of his acceleration, including both the time when he is in the air and the time when his feet are in contact with the trampoline.

B. Sally is on an amusement park ride which begins with her chair being hoisted straight up a tower at a constant speed of 60 miles/hour. Despite stern warnings from her father that he'll take her home the next time she misbehaves, she decides that as a scientific experiment she really needs to release her corndog over the side as she's on the way up. She does not throw it. She simply sticks it out of the car, lets it go, and watches it against the background of the sky, with no trees or buildings as reference points. What does the corndog's motion look like as observed by Sally? Does its speed ever appear to her to be zero? What acceleration does she observe it to have: is it ever positive? negative? zero? What would her enraged father answer if asked for a similar description of its motion as it appears to him, standing on the ground?

C. Can an object maintain a constant acceleration, but meanwhile reverse the direction of its velocity?

D. Can an object have a velocity that is positive and increasing at the same time that its acceleration is decreasing?

E. The four figures show a refugee from a Picasso painting blowing on a rolling water bottle. In some cases the person's blowing is speeding the bottle up, but in others it is slowing it down. The arrow inside the bottle shows which

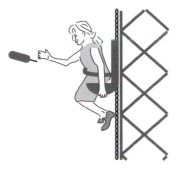

Discussion question B.

direction it is going, and a coordinate system is shown at the bottom of each figure. In each case, figure out the plus or minus signs of the velocity and acceleration. It may be helpful to draw a *v-t* graph in each case.

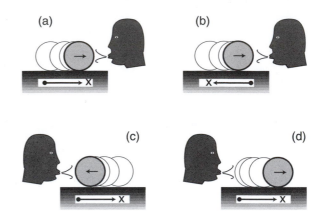

3.4 Varying Acceleration

So far we have only been discussing examples of motion for which the *v-t* graph is linear. If we wish to generalize our definition to v-t graphs that are more complex curves, the best way to proceed is similar to how we defined velocity for curved *x-t* graphs:

definition of acceleration
The acceleration of an object at any instant is the slope of the tangent line passing through its *v*-versus-*t* graph at the relevant point.

Example: a skydiver
Question: The graphs show the results of a fairly realistic computer simulation of the motion of a skydiver, including the effects of air friction. The *x* axis has been chosen pointing down, so *x* is increasing as she falls. Find (a) the skydiver's acceleration at t=3.0 s, and also (b) at *t*=7.0 s.
Solution: I've added tangent lines at the two points in question. (a) To find the slope of the tangent line, I pick two points on the

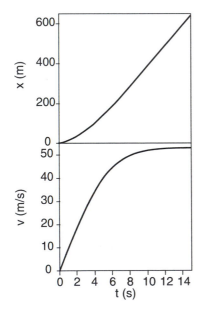

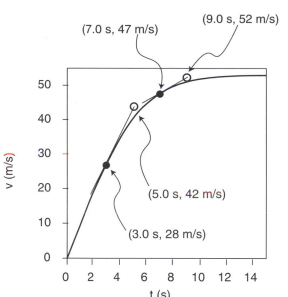

(7.0 s, 47 m/s)

(9.0 s, 52 m/s)

(5.0 s, 42 m/s)

(3.0 s, 28 m/s)

line (not necessarily on the actual curve): (3.0 s, 28 m/s) and (5.0 s, 42 m/s). The slope of the tangent line is (42 m/s-28 m/s)/(5.0 s - 3.0 s)=7.0 m/s².

(b) Two points on this tangent line are (7.0 s, 47 m/s) and (9.0 s, 52 m/s). The slope of the tangent line is (52 m/s-47 m/s)/(9.0 s - 7.0 s)=2.5 m/s².

Physically, what's happening is that at t=3.0 s, the skydiver is not yet going very fast, so air friction is not yet very strong. She therefore has an acceleration almost as great as g. At t=7.0 s, she is moving almost twice as fast (about 100 miles per hour), and air friction is extremely strong, resulting in a significant departure from the idealized case of no air friction.

In the above example, the *x-t* graph was not even used in the solution of the problem, since the definition of acceleration refers to the slope of the *v-t* graph. It is possible, however, to interpret an *x-t* graph to find out something about the acceleration. An object with zero acceleration, i.e. constant velocity, has an *x-t* graph that is a straight line. A straight line has no curvature. A change in velocity requires a change in the slope of the x-t graph, which means that it is a curve rather than a line. Thus acceleration relates to the curvature of the *x-t* graph. Figure (c) shows some examples.

In the skydiver example, the *x-t* graph was more strongly curved at the beginning, and became nearly straight at the end. If the *x-t* graph is nearly straight, then its slope, the velocity, is nearly constant, and the acceleration is therefore small. We can thus interpret the acceleration as representing the curvature of the *x-t* graph. If the "cup" of the curve points up, the acceleration is positive, and if it points down, the acceleration is negative.

Since the relationship between *a* and *v* is analogous to the relationship

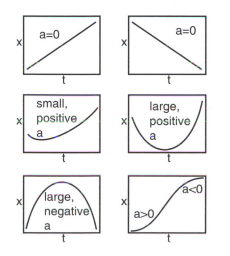

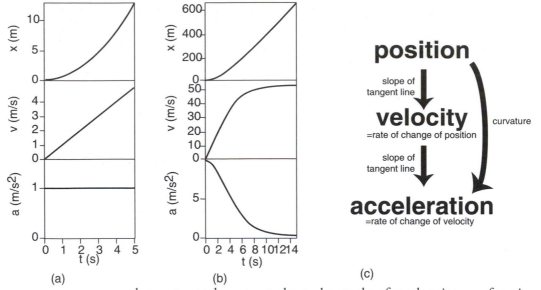

(a)   (b)   (c)

between v and x, we can also make graphs of acceleration as a function of time, as shown in figures (a) and (b) above.

Figure (c) summarizes the relationships among the three types of graphs.

Discussion questions

A. Describe in words how the changes in the a-t graph for the skydiver relate to the behavior of the v-t graph.

B. Explain how each set of graphs contains inconsistencies.

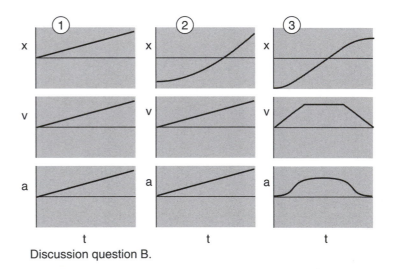

Discussion question B.

C. In each case, pick a coordinate system and draw x-t, v-t, and a-t graphs. Picking a coordinate system means picking where you want x=0 to be, and also picking a direction for the positive x axis.

1. An ocean liner is crusing in a straight line at constant speed.

2. You drop a ball. Draw to different sets of graphs (a total of 6), with one set's positive x axis pointing in the opposite direction compared to the other's.

3. You're driving down the street looking for a house you've never been to before. You realize you've passed the address, so you slow down, put the car in reverse, back up, and stop in front of the house.

3.5 The Area Under the Velocity-Time Graph

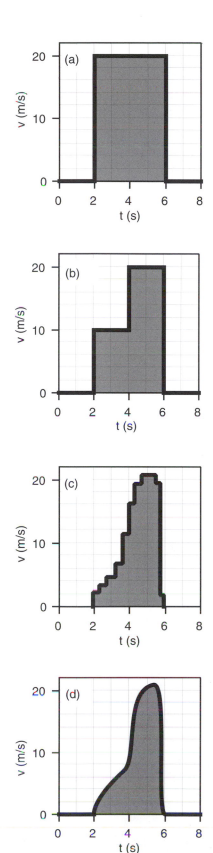

A natural question to ask about falling objects is how fast they fall, but Galileo showed that the question has no answer. The physical law that he discovered connects a cause (the attraction of the planet Earth's mass) to an effect, but the effect is predicted in terms of an acceleration rather than a velocity. In fact, no physical law predicts a definite velocity as a result of a specific phenomenon, because velocity cannot be measured in absolute terms, and only changes in velocity relate directly to physical phenomena.

The unfortunate thing about this situation is that the definitions of velocity and acceleration are stated in terms of the tangent-line technique, which lets you go from x to v to a, but not the other way around. Without a technique to go backwards from a to v to x, we cannot say anything quantitative, for instance, about the x-t graph of a falling object. Such a technique does exist, and I used it to make the x-t graphs in all the examples above.

First let's concentrate on how to get x information out of a v-t graph. In example (a), an object moves at a speed of 20 m/s for a period of 4.0 s. The distance covered is $\Delta x = v \Delta t = (20 \text{ m/s}) \times (4.0 \text{ s}) = 80 \text{ m}$. Notice that the quantities being multiplied are the width and the height of the shaded rectangle — or, strictly speaking, the time represented by its width and the velocity represented by its height. The distance of $\Delta x = 80$ m thus corresponds to the area of the shaded part of the graph.

The next step in sophistication is an example like (b), where the object moves at a constant speed of 10 m/s for two seconds, then for two seconds at a different constant speed of 20 m/s. The shaded region can be split into a small rectangle on the left, with an area representing $\Delta x = 20$ m, and a taller one on the right, corresponding to another 40 m of motion. The total distance is thus 60 m, which corresponds to the total area under the graph.

An example like (c) is now just a trivial generalization; there is simply a large number of skinny rectangular areas to add up. But notice that graph (c) is quite a good approximation to the smooth curve (d). Even though we have no formula for the area of a funny shape like (d), we can approximate its area by dividing it up into smaller areas like rectangles, whose area is easier to calculate. If someone hands you a graph like (d) and asks you to find the area under it, the simplest approach is just to count up the little rectangles on the underlying graph paper, making rough estimates of fractional rectangles as you go along.

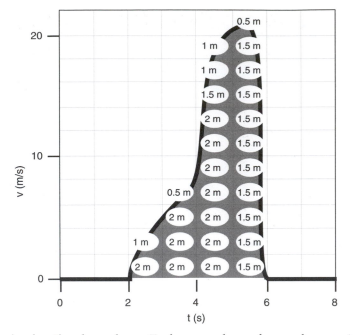

That's what I've done above. Each rectangle on the graph paper is 1.0 s wide and 2 m/s tall, so it represents 2 m. Adding up all the numbers gives $\Delta x = 41$ m. If you needed better accuracy, you could use graph paper with smaller rectangles.

It's important to realize that this technique gives you Δx, not x. The v-t graph has no information about where the object was when it started.

The following are important points to keep in mind when applying this technique:

- If the range of v values on your graph does not extend down to zero, then you will get the wrong answer unless you compensate by adding in the area that is not shown.

- As in the example, one rectangle on the graph paper does not necessarily correspond to one meter of distance.

- Negative velocity values represent motion in the opposite direction, so area under the t axis should be subtracted, i.e. counted as "negative area."

- Since the result is a Δx value, it only tells you x_{after}-x_{before}, which may be less than the actual distance traveled. For instance, the object could come back to its original position at the end, which would correspond to $\Delta x = 0$, even though it had actually moved a nonzero distance.

Finally, note that one can find Δv from an a-t graph using an entirely analogous method. Each rectangle on the a-t graph represents a certain amount of velocity change.

Discussion question

Roughly what would a pendulum's v-t graph look like? What would happen when you applied the area-under-the-curve technique to find the pendulum's Δx for a time period covering many swings?

3.6 Algebraic Results for Constant Acceleration

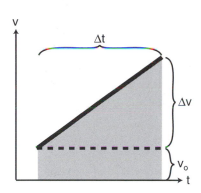

Although the area-under-the-curve technique can be applied to any graph, no matter how complicated, it may be laborious to carry out, and if fractions of rectangles must be estimated the result will only be approximate. In the special case of motion with constant acceleration, it is possible to find a convenient shortcut which produces exact results. When the acceleration is constant, the $v\text{-}t$ graph is a straight line, as shown in the figure. The area under the curve can be divided into a triangle plus a rectangle, both of whose areas can be calculated exactly: $A=bh$ for a rectangle and $A=\frac{1}{2}bh$ for a triangle. The height of the rectangle is the initial velocity, v_o, and the height of the triangle is the change in velocity from beginning to end, Δv. The object's Δx is therefore given by the equation

$\Delta x = v_o\Delta t + \frac{1}{2}\Delta v\Delta t$. This can be simplified a little by using the definition of acceleration, $a=\Delta v/\Delta t$ to eliminate Δv, giving

$$\Delta x = v_o\Delta t + \frac{1}{2}a\Delta t^2 \quad \text{[motion with constant acceleration]} .$$

Since this is a second-order polynomial in Δt, the graph of Δx versus Δt is a parabola, and the same is true of a graph of x versus t — the two graphs differ only by shifting along the two axes. Although I have derived the equation using a figure that shows a positive v_o, positive a, and so on, it still turns out to be true regardless of what plus and minus signs are involved.

Another useful equation can be derived if one wants to relate the change in velocity to the distance traveled. This is useful, for instance, for finding the distance needed by a car to come to a stop. For simplicity, we start by deriving the equation for the special case of $v_o=0$, in which the final velocity v_f is a synonym for Δv. Since velocity and distance are the variables of interest, not time, we take the equation $\Delta x = \frac{1}{2}a\Delta t^2$ and use $\Delta t=\Delta v/a$ to eliminate Δt. This gives $\Delta x = \frac{1}{2}(\Delta v)^2/a$, which can be rewritten as

$$v_f^2 = 2a\Delta x \quad \text{[motion with constant acceleration, } v_o = 0] .$$

For the more general case where $v_o \neq 0$, we skip the tedious algebra leading to the more general equation,

$$v_f^2 = v_o^2 + 2a\Delta x \quad \text{[motion with constant acceleration]} .$$

To help get this all organized in your head, first let's categorize the variables as follows:

Variables that change during motion with constant acceleration:

x, v, t

Variable that doesn't change:

a

If you know one of the changing variables and want to find another, there is always an equation that relates those two:

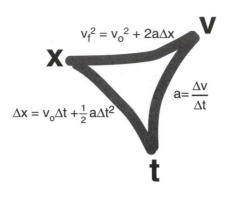

The symmetry among the three variables is imperfect only because the equation relating x and t includes the initial velocity.

There are two main difficulties encountered by students in applying these equations:

- The equations apply only to motion with constant acceleration. You can't apply them if the acceleration is changing.

- Students are often unsure of which equation to use, or may cause themselves unnecessary work by taking the longer path around the triangle in the chart above. Organize your thoughts by listing the variables you are given, the ones you want to find, and the ones you aren't given and don't care about.

Example
Question: You are trying to pull an old lady out of the way of an oncoming truck. You are able to give her an acceleration of 20 m/s^2. Starting from rest, how much time is required in order to move her 2 m?
Solution: First we organize our thoughts:

Variables given:	$\Delta x, a, v_o$
Variables desired:	Δt
Irrelevant variables:	v_f

Consulting the triangular chart above, the equation we need is clearly $\Delta x = v_o \Delta t + \frac{1}{2} a \Delta t^2$, since it has the four variables of interest and omits the irrelevant one. Eliminating the v_o term and solving for Δt gives $\Delta t = \sqrt{\frac{2 \Delta x}{a}}$ =0.4 s.

Discussion questions

A Check that the units make sense in the three equations derived in this section.
B. In chapter 1, I gave examples of correct and incorrect reasoning about proportionality, using questions about the scaling of area and volume. Try to translate the incorrect modes of reasoning shown there into mistakes about the following question: If the acceleration of gravity on Mars is 1/3 that on Earth, how many times longer does it take for a rock to drop the same distance on Mars?

3.7* Biological Effects of Weightlessness

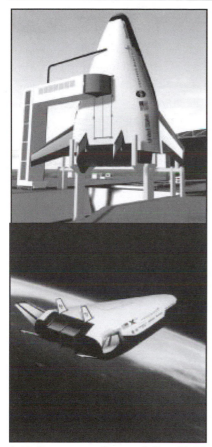

Artist's conceptions of the X-33 spaceship, a half-scale uncrewed version of the planned VentureStar vehicle, which was supposed to cut the cost of sending people into space by an order of magnitude. The X-33 program was canceled in March 2001 due to technical failures and budget overruns, so the Space Shuttle will remain the U.S.'s only method of sending people into space for the forseeable future.
Courtesy of NASA.

The usefulness of outer space was brought home to North Americans in 1998 by the unexpected failure of the communications satellite that had been handling almost all of the continent's cellular phone traffic. Compared to the massive economic and scientific payoffs of satellites and space probes, human space travel has little to boast about after four decades. Sending people into orbit has just been too expensive to be an effective scientific or commercial activity. The 1986 Challenger disaster dealt a blow to NASA's confidence, and with the end of the cold war, U.S. prestige as a superpower was no longer a compelling reason to send Americans into space. All that may change soon, with a new generation of much cheaper reusable space-ships. (The space shuttle is not truly reusable. Retrieving the boosters out of the ocean is no cheaper than building new ones, but NASA brings them back and uses them over for public relations, to show how frugal they are.) Space tourism is even beginning to make economic sense! No fewer than three private companies are now willing to take your money for a reservation on a two-to-four minute trip into space, although none of them has a firm date on which to begin service. Within a decade, a space cruise may be the new status symbol among those sufficiently rich and brave.

Space sickness

Well, rich, brave, and possessed of an iron stomach. Travel agents will probably not emphasize the certainty of constant space-sickness. For us animals evolved to function in g=9.8 m/s^2, living in g=0 is extremely unpleasant. The early space program focused obsessively on keeping the astronaut-trainees in perfect physical shape, but it soon became clear that a body like a Greek demigod's was no defense against that horrible feeling that your stomach was falling out from under you and you were never going to catch up. Our inner ear, which normally tells us which way is down, tortures us when down is nowhere to be found. There is contradictory information about whether anyone ever gets over it; the "right stuff" culture creates a strong incentive for astronauts to deny that they are sick.

Effects of long space missions

Worse than nausea are the health-threatening effects of prolonged weightlessness. The Russians are the specialists in long-term missions, in which cosmonauts suffer harm to their blood, muscles, and, most importantly, their bones.

The effects on the muscles and skeleton appear to be similar to those experienced by old people and people confined to bed for a long time. Everyone knows that our muscles get stronger or weaker depending on the amount of exercise we get, but the bones are likewise adaptable. Normally old bone mass is continually being broken down and replaced with new material, but the balance between its loss and replacement is upset when people do not get enough weight-bearing exercise. The main effect is on the bones of the lower body. More research is required to find out whether astronauts' loss of bone mass is due to faster breaking down of bone, slower replacement, or both. It is also not known whether the effect can be suppressed via diet or drugs.

The other set of harmful physiological effects appears to derive from the redistribution of fluids. Normally, the veins and arteries of the legs are

U.S. and Russian astronauts aboard the International Space Station, October 2000.

The International Space Station, September 2000. The space station will not rotate to provide simulated gravity. The completed station will be much bigger than it is in this picture.

More on Apparent Weightlessness
Astronauts in orbit are not really weightless; they are only a few hundred miles up, so they are still affected strongly by the Earth's gravity. Section 10.3 of this book discusses why they experience apparent weightlessness.

More on Simulated Gravity
For more information on simulating gravity by spinning a spacecraft, see section 9.2 of this book.

tightly constricted to keep gravity from making blood collect there. It is uncomfortable for adults to stand on their heads for very long, because the head's blood vessels are not able to constrict as effectively. Weightless astronauts' blood tends to be expelled by the constricted blood vessels of the lower body, and pools around their hearts, in their thoraxes, and in their heads. The only immediate result is an uncomfortable feeling of bloatedness in the upper body, but in the long term, a harmful chain of events is set in motion. The body's attempts to maintain the correct blood volume are most sensitive to the level of fluid in the head. Since astronauts have extra fluid in their heads, the body thinks that the over-all blood volume has become too great. It responds by decreasing blood volume below normal levels. This increases the concentration of red blood cells, so the body then decides that the blood has become too thick, and reduces the number of blood cells. In missions lasting up to a year or so, this is not as harmful as the musculo-skeletal effects, but it is not known whether longer period in space would bring the red blood cell count down to harmful levels.

Reproduction in space

For those enthralled by the romance of actual human colonization of space, human reproduction in weightlessness becomes an issue. An already-pregnant Russian cosmonaut did spend some time in orbit in the 1960's, and later gave birth to a normal child on the ground. Recently, one of NASA's public relations concerns about the space shuttle program has been to discourage speculation about space sex, for fear of a potential taxpayers' backlash against the space program as an expensive form of exotic pleasure.

Scientific work has been concentrated on studying plant and animal reproduction in space. Green plants, fungi, insects, fish, and amphibians have all gone through at least one generation in zero-gravity experiments without any serious problems. In many cases, animal embryos conceived in orbit begin by developing abnormally, but later in development they seem to correct themselves. However, chicken embryos fertilized on earth less than 24 hours before going into orbit have failed to survive. Since chickens are the organisms most similar to humans among the species investigated so far, it is not at all certain that humans could reproduce successfully in a zero-gravity space colony.

Simulated gravity

If humans are ever to live and work in space for more than a year or so, the only solution is probably to build spinning space stations to provide the illusion of weight, as discussed in section 9.2. Normal gravity could be simulated, but tourists would probably enjoy $g=2$ m/s^2 or 5 m/s^2. Space enthusiasts have proposed entire orbiting cities built on the rotating cylinder plan. Although science fiction has focused on human colonization of relatively earthlike bodies such as our moon, Mars, and Jupiter's icy moon Europa, there would probably be no practical way to build large spinning structures on their surfaces. If the biological effects of their 2-3 m/s^2 gravitational accelerations are as harmful as the effect of $g=0$, then we may be left with the surprising result that interplanetary space is more hospitable to our species than the moons and planets.

3.8 ∫ Applications of Calculus

In the Applications of Calculus section at the end of the previous chapter, I discussed how the slope-of-the-tangent-line idea related to the calculus concept of a derivative, and the branch of calculus known as differential calculus. The other main branch of calculus, integral calculus, has to do with the area-under-the-curve concept discussed in section 3.5 of this chapter. Again there is a concept, a notation, and a bag of tricks for doing things symbolically rather than graphically. In calculus, the area under the v-t graph between $t=t_1$ and $t=t_2$ is notated like this:

$$\text{area under the curve} = \Delta x = \int_{t_1}^{t_2} v \, dt$$

The expression on the right is called an integral, and the s-shaped symbol, the integral sign, is read as "integral of...."

Integral calculus and differential calculus are closely related. For instance, if you take the derivative of the function $x(t)$, you get the function $v(t)$, and if you integrate the function $v(t)$, you get $x(t)$ back again. In other words, integration and differentiation are inverse operations. This is known as the fundamental theorem of calculus.

On an unrelated topic, there is a special notation for taking the derivative of a function twice. The acceleration, for instance, is the second (i.e. double) derivative of the position, because differentiating x once gives v, and then differentiating v gives a. This is written as

$$a = \frac{d^2 x}{dt^2} \quad .$$

The seemingly inconsistent placement of the twos on the top and bottom confuses all beginning calculus students. The motivation for this funny notation is that acceleration has units of m/s^2, and the notation correctly suggests that: the top looks like it has units of meters, the bottom seconds^2. The notation is not meant, however, to suggest that t is really squared.

Summary

Selected Vocabulary

gravity .... A general term for the phenomenon of attraction between things having mass. The attraction between our planet and a human-sized object causes the object to fall.

acceleration .... The rate of change of velocity; the slope of the tangent line on a *v-t* graph.

Notation

a .... acceleration

g .... the acceleration of objects in free fall

Summary

Galileo showed that when air resistance is negligible all falling bodies have the same motion regardless of mass. Moreover, their *v-t* graphs are straight lines. We therefore define a quantity called acceleration as the slope, $\Delta v/\Delta t$, of an object's *v-t* graph. In cases other than free fall, the *v-t* graph may be curved, in which case the definition is generalized as the slope of a tangent line on the *v-t* graph. The acceleration of objects in free fall varies slightly across the surface of the earth, and greatly on other planets.

Positive and negative signs of acceleration are defined according to whether the *v-t* graph slopes up or down. This definition has the advantage that a force in a given direction always produces the same sign of acceleration.

The area under the *v-t* graph gives Δx, and analogously the area under the *a-t* graph gives Δv.

For motion with constant acceleration, the following three equations hold:

$$\Delta x = v_o\Delta t + \tfrac{1}{2}a\Delta t^2$$

$$v_f^2 = v_o^2 + 2a\Delta x$$

$$a = \frac{\Delta v}{\Delta t}$$

They are not valid if the acceleration is changing.

Homework Problems

1 ✓. The graph represents the velocity of a bee along a straight line. At $t=0$, the bee is at the hive. (a) When is the bee farthest from the hive? (b) How far is the bee at its farthest point from the hive? (c) At $t=13$ s, how far is the bee from the hive? [Hint: Try problem 19 first.]

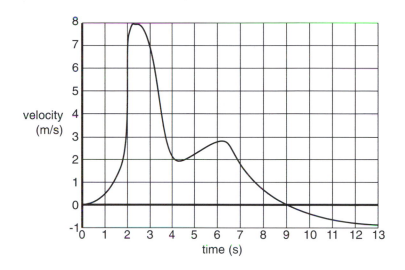

2. A rock is dropped into a pond. Draw plots of its position versus time, velocity versus time, and acceleration versus time. Include its whole motion, starting from the moment it is dropped, and continuing while it falls through the air, passes through the water, and ends up at rest on the bottom of the pond.

3. In an 18th-century naval battle, a cannon ball is shot horizontally, passes through the side of an enemy ship's hull, flies across the galley, and lodges in a bulkhead. Draw plots of its horizontal position, velocity, and acceleration as functions of time, starting while it is inside the cannon and has not yet been fired, and ending when it comes to rest. There is not any significant amount of friction from the air. Although the ball may rise and fall, you are only concerned with its horizontal motion, as seen from above.

4. Draw graphs of position, velocity, and acceleration as functions of time for a person bunjee jumping. (In bunjee jumping, a person has a stretchy elastic cord tied to his/her ankles, and jumps off of a high platform. At the bottom of the fall, the cord brings the person up short. Presumably the person bounces up a little.)

5. A ball rolls down the ramp shown in the figure, consisting of a curved knee, a straight slope, and a curved bottom. For each part of the ramp, tell whether the ball's velocity is increasing, decreasing, or constant, and also whether the ball's acceleration is increasing, decreasing, or constant. Explain your answers. Assume there is no air friction or rolling resistance. Hint: Try problem 20 first. [Based on a problem by Hewitt.]

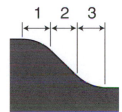

Problem 3.

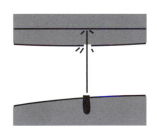

Problem 5.

6. A toy car is released on one side of a piece of track that is bent into an upright U shape. The car goes back and forth. When the car reaches the limit of its motion on one side, its velocity is zero. Is its acceleration also zero? Explain using a *v-t* graph. [Based on a problem by Serway and Faughn.]

7. What is the acceleration of a car that moves at a steady velocity of 100 km/h for 100 seconds? Explain your answer. [Based on a problem by Hewitt.]

8. A physics homework question asks, "If you start from rest and accelerate at 1.54 m/s^2 for 3.29 s, how far do you travel by the end of that time?" A student answers as follows:

$$1.54 \times 3.29 = 5.07 \text{ m}$$

His Aunt Wanda is good with numbers, but has never taken physics. She doesn't know the formula for the distance traveled under constant acceleration over a given amount of time, but she tells her nephew his answer cannot be right. How does she know?

9 ✓. You are looking into a deep well. It is dark, and you cannot see the bottom. You want to find out how deep it is, so you drop a rock in, and you hear a splash 3.0 seconds later. How deep is the well?

10 ★✓. You take a trip in your spaceship to another star. Setting off, you increase your speed at a constant acceleration. Once you get half-way there, you start decelerating, at the same rate, so that by the time you get there, you have slowed down to zero speed. You see the tourist attractions, and then head home by the same method.

(a) Find a formula for the time, *T*, required for the round trip, in terms of *d*, the distance from our sun to the star, and *a*, the magnitude of the acceleration. Note that the acceleration is not constant over the whole trip, but the trip can be broken up into constant-acceleration parts.

(b) The nearest star to the Earth (other than our own sun) is Proxima Centauri, at a distance of $d=4\times10^{16}$ m. Suppose you use an acceleration of $a=10$ m/s^2, just enough to compensate for the lack of true gravity and make you feel comfortable. How long does the round trip take, in years?

(c) Using the same numbers for *d* and *a*, find your maximum speed. Compare this to the speed of light, which is 3.0×10^8 m/s. (Later in this course, you will learn that there are some new things going on in physics when one gets close to the speed of light, and that it is impossible to exceed the speed of light. For now, though, just use the simpler ideas you've learned so far.)

11. You climb half-way up a tree, and drop a rock. Then you climb to the top, and drop another rock. How many times greater is the velocity of the second rock on impact? Explain. (The answer is not two times greater.)

12. Alice drops a rock off a cliff. Bubba shoots a gun straight down from the edge of the same cliff. Compare the accelerations of the rock and the bullet while they are on the way down. [Based on a problem by Serway and Faughn.]

13 $\int$. A person is parachute jumping. During the time between when she leaps out of the plane and when she opens her chute, her altitude is given by an equation of the form

$$y = b - c\left(t + ke^{-t/k}\right) \quad ,$$

where e is the base of natural logarithms, and b, c, and k are constants. Because of air resistance, her velocity does not increase at a steady rate as it would for an object falling in vacuum.

(a) What units would b, c, and k have to have for the equation to make sense?

(b✓) Find the person's velocity, v, as a function of time. [You will need to use the chain rule, and the fact that $d(e^x)/dx=e^x$.]

(c) Use your answer from part (b) to get an interpretation of the constant c. [Hint: e^{-x} approaches zero for large values of x.]

(d✓) Find the person's acceleration, a, as a function of time.

(e) Use your answer from part (b) to show that if she waits long enough to open her chute, her acceleration will become very small.

14 S. The top part of the figure shows the position-versus-time graph for an object moving in one dimension. On the bottom part of the figure, sketch the corresponding v-versus-t graph.

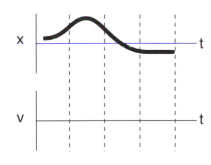

Problem 14.

15 S. On New Year's Eve, a stupid person fires a pistol straight up. The bullet leaves the gun at a speed of 100 m/s. How long does it take before the bullet hits the ground?

16 S. If the acceleration of gravity on Mars is 1/3 that on Earth, how many times longer does it take for a rock to drop the same distance on Mars? Ignore air resistance.

17 S$\int$. A honeybee's position as a function of time is given by $x=10t-t^3$, where t is in seconds and x in meters. What is its acceleration at $t=3.0$ s?

18 S. In July 1999, Popular Mechanics carried out tests to find which car sold by a major auto maker could cover a quarter mile (402 meters) in the shortest time, starting from rest. Because the distance is so short, this type of test is designed mainly to favor the car with the greatest acceleration, not the greatest maximum speed (which is irrelevant to the average person). The winner was the Dodge Viper, with a time of 12.08 s. The car's top (and presumably final) speed was 118.51 miles per hour (52.98 m/s). (a) If a car, starting from rest and moving with *constant* acceleration, covers a quarter mile in this time interval, what is its acceleration? (b) What would be the final speed of a car that covered a quarter mile with the constant acceleration you found in part a? (c) Based on the discrepancy between your answer in part b and the actual final speed of the Viper, what do you conclude about how its acceleration changed over time?

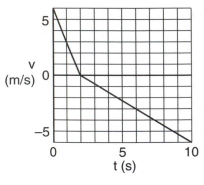

Problem 19.

Problem 20.

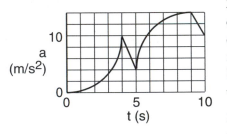

Problem 23.

19 S. The graph represents the motion of a rolling ball that bounces off of a wall. When does the ball return to the location it had at $t=0$?

20 S. (a) The ball is released at the top of the ramp shown in the figure. Friction is negligible. Use physical reasoning to draw v-t and a-t graphs. Assume that the ball doesn't bounce at the point where the ramp changes slope. (b) Do the same for the case where the ball is rolled up the slope from the right side, but doesn't quite have enough speed to make it over the top.

21 S. You throw a rubber ball up, and it falls and bounces several times. Draw graphs of position, velocity, and acceleration as functions of time.

22 S. Starting from rest, a ball rolls down a ramp, traveling a distance L and picking up a final speed v. How much of the distance did the ball have to cover before achieving a speed of $v/2$? [Based on a problem by Arnold Arons.]

23✓. The graph shows the acceleration of a chipmunk in a TV cartoon. It consists of two circular arcs and two line segments. At $t=0$, the chipmunk's velocity is -3.1 m/s. What is its velocity at $t=10$ s?

24. Find the error in the following calculation. A student wants to find the distance traveled by a car that accelerates from rest for 5.0 s with an acceleration of 2.0 m/s². First he solves $a=\Delta v/\Delta t$ for $\Delta v=10$ m/s. Then he multiplies to find $(10 \text{ m/s})(5.0 \text{ s})=50$ m. Do not just recalculate the result by a different method; if that was all you did, you'd have no way of knowing which calculation was correct, yours or his.

25. Acceleration could be defined either as $\Delta v/\Delta t$ or as the slope of the tangent line on the v-t graph. Is either one superior as a definition, or are they equivalent? If you say one is better, give an example of a situation where it makes a difference which one you use.

26. If an object starts accelerating from rest, we have $v^2=2a\Delta x$ for its speed after it has traveled a distance Δx. Explain in words why it makes sense that the equation has velocity squared, but distance only to the first power. Don't recapitulate the derivation in the book, or give a justification based on units. The point is to explain what this feature of the equation tells us about the way speed increases as more distance is covered.

27✓. The figure shows a practical, simple experiment for determining g to high precision. Two steel balls are suspended from electromagnets, and are released simultaneously when the electric current is shut off. They fall through unequal heights Δx_1 and Δx_2. A computer records the sounds through a microphone as first one ball and then the other strikes the floor. From this recording, we can accurately determine the quantity T defined as $T=\Delta t_2-\Delta t_1$, i.e., the time lag between the first and second impacts. Note that since the balls do not make any sound when they are released, we have no way of measuring the individual times Δt_2 and Δt_1. (a✓) Find an equation for g in terms of the measured quantities T, Δx_1 and Δx_2. (b) Check the units of your equation. (c) Check that your equation gives the correct result in the case where $\Delta x_1=0$. However, is this case realistic? (d) What happens when $\Delta x_1=\Delta x_2$?

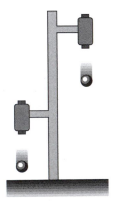

Problem 27.

Isaac Newton

Even as great and skeptical a genius as Galileo was unable to make much progress on the causes of motion. It was not until a generation later that Isaac Newton (1642-1727) was able to attack the problem successfully. In many ways, Newton's personality was the opposite of Galileo's. Where Galileo agressively publicized his ideas, Newton had to be coaxed by his friends into publishing a book on his physical discoveries. Where Galileo's writing had been popular and dramatic, Newton originated the stilted, impersonal style that most people think is standard for scientific writing. (Scientific journals today encourage a less ponderous style, and papers are often written in the first person.) Galileo's talent for arousing animosity among the rich and powerful was matched by Newton's skill at making himself a popular visitor at court. Galileo narrowly escaped being burned at the stake, while Newton had the good fortune of being on the winning side of the revolution that replaced King James II with William and Mary of Orange, leading to a lucrative post running the English royal mint.

Newton discovered the relationship between force and motion, and revolutionized our view of the universe by showing that the same physical laws applied to all matter, whether living or nonliving, on or off of our planet's surface. His book on force and motion, the **Mathematical Principles of Natural Philosophy**, was uncontradicted by experiment for 200 years, but his other main work, **Optics**, was on the wrong track due to his conviction that light was composed of particles rather than waves. Newton was also an avid alchemist, an embarrassing fact that modern scientists would like to forget.

4 Force and Motion

If I have seen farther than others, it is because I have stood on the shoulders of giants.

Newton, referring to Galileo

4.1 Force

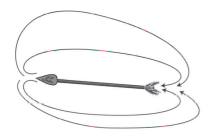

Aristotle said motion had to be caused by a force. To explain why an arrow kept flying after the bowstring was no longer pushing on it, he said the air rushed around behind the arrow and pushed it forward. We know this is wrong, because an arrow shot in a vacuum chamber does not instantly drop to the floor as it leaves the bow. Galileo and Newton realized that a force would only be needed to change the arrow's motion, not to make its motion continue.

We need only explain changes in motion, not motion itself

So far you've studied the measurement of motion in some detail, but not the reasons why a certain object would move in a certain way. This chapter deals with the "why" questions. Aristotle's ideas about the causes of motion were completely wrong, just like all his other ideas about physical science, but it will be instructive to start with them, because they amount to a road map of modern students' incorrect preconceptions.

Aristotle thought he needed to explain both why motion occurs and why motion might change. Newton inherited from Galileo the important counter-Aristotelian idea that motion needs no explanation, that it is only *changes* in motion that require a physical cause.

Aristotle gave three reasons for motion:

- Natural motion, such as falling, came from the tendency of objects to go to their "natural" place, on the ground, and come to rest.

- Voluntary motion was the type of motion exhibited by animals, which moved because they chose to.

- Forced motion occurred when an object was acted on by some other object that made it move.

Motion changes due to an interaction between two objects

In the Aristotelian theory, natural motion and voluntary motion are one-sided phenomena: the object causes its own motion. Forced motion is supposed to be a two-sided phenomenon, because one object imposes its "commands" on another. Where Aristotle conceived of some of the phenomena of motion as one-sided and others as two-sided, Newton realized that a change in motion was always a two-sided relationship of a force acting between two physical objects.

The one-sided "natural motion" description of falling makes a crucial omission. The acceleration of a falling object is not caused by its own "natural" tendencies but by an attractive force between it and the planet Earth. Moon rocks brought back to our planet do not "want" to fly back up to the moon because the moon is their "natural" place. They fall to the floor when you drop them, just like our homegrown rocks. As we'll discuss in more detail later in this course, gravitational forces are simply an attraction that occurs between any two physical objects. Minute gravitational forces can even be measured between human-scale objects in the laboratory.

The idea of natural motion also explains incorrectly why things come to rest. A basketball rolling across a beach slows to a stop because it is interacting with the sand via a frictional force, not because of its own desire to be at rest. If it was on a frictionless surface, it would never slow down. Many of Aristotle's mistakes stemmed from his failure to recognize friction as a force.

"Our eyes receive blue light reflected from this painting because Monet wanted to represent water with the color blue." This is a valid statement at one level of explanation, but physics works at the physical level of explanation, in which blue light gets to your eyes because it is reflected by blue pigments in the paint.

The concept of voluntary motion is equally flawed. You may have been a little uneasy about it from the start, because it assumes a clear distinction between living and nonliving things. Today, however, we are used to having the human body likened to a complex machine. In the modern world-view, the border between the living and the inanimate is a fuzzy no-man's land inhabited by viruses, prions, and silicon chips. Furthermore, Aristotle's statement that you can take a step forward "because you choose to" inappropriately mixes two levels of explanation. At the physical level of explanation, the reason your body steps forward is because of a frictional force acting between your foot and the floor. If the floor was covered with a puddle of oil, no amount of "choosing to" would enable you to take a graceful stride forward.

Forces can all be measured on the same numerical scale

In the Aristotelian-scholastic tradition, the description of motion as natural, voluntary, or forced was only the broadest level of classification, like splitting animals into birds, reptiles, mammals, and amphibians. There might be thousands of types of motion, each of which would follow its own rules. Newton's realization that all changes in motion were caused by two-sided interactions made it seem that the phenomena might have more in common than had been apparent. In the Newtonian description, there is only one cause for a change in motion, which we call force. Forces may be of different types, but they all produce changes in motion according to the same rules. Any acceleration that can be produced by a magnetic force can equally well be produced by an appropriately controlled stream of water. We can speak of two forces as being equal if they produce the same change in motion when applied in the same situation, which means that they pushed or pulled equally hard in the same direction.

The idea of a numerical scale of force and the newton unit were introduced in chapter 0. To recapitulate briefly, a force is when a pair of objects push or pull on each other, and one newton is the force required to accelerate a 1-kg object from rest to a speed of 1 m/s in 1 second.

More than one force on an object

As if we hadn't kicked poor Aristotle around sufficiently, his theory has another important flaw, which is important to discuss because it corresponds to an extremely common student misconception. Aristotle conceived of forced motion as a relationship in which one object was the boss and the other "followed orders." It therefore would only make sense for an object to experience one force at a time, because an object couldn't follow orders from two sources at once. In the Newtonian theory, forces are numbers, not orders, and if more than one force acts on an object at once, the result is found by adding up all the forces. It is unfortunate that the use the English word "force" has become standard, because to many people it suggests that you are "forcing" an object to do something. The force of the earth's gravity cannot "force" a boat to sink, because there are other forces acting on the boat. Adding them up gives a total of zero, so the boat accelerates neither up nor down.

Objects can exert forces on each other at a distance

Aristotle declared that forces could only act between objects that were touching, probably because he wished to avoid the type of occult speculation that attributed physical phenomena to the influence of a distant and invisible pantheon of gods. He was wrong, however, as you can observe when a magnet leaps onto your refrigerator or when the planet earth exerts gravitational forces on objects that are in the air. Some types of forces, such as friction, only operate between objects in contact, and are called *contact forces*. Magnetism, on the other hand, is an example of a *noncontact force*. Although the magnetic force gets stronger when the magnet is closer to your refrigerator, touching is not required.

Weight

In physics, an object's weight , F_W, is defined as the earth's gravitational force on it. The SI unit of weight is therefore the Newton. People commonly refer to the kilogram as a unit of weight, but the kilogram is a unit of mass, not weight. Note that an object's weight is not a fixed property of that object. Objects weigh more in some places than in others, depending on the local strength of gravity. It is their mass that always stays the same. A baseball pitcher who can throw a 90-mile-per-hour fastball on earth would not be able to throw any faster on the moon, because the ball's inertia would still be the same.

Positive and negative signs of force

We'll start by considering only cases of one-dimensional center-of-mass motion in which all the forces are parallel to the direction of motion, i.e. either directly forward or backward. In one dimension, plus and minus signs can be used to indicate directions of forces, as shown in the figure. We can then refer generically to addition of forces, rather than having to speak sometimes of addition and sometimes of subtraction. We add the forces shown in the figure and get 11 N. In general, we should choose a one-

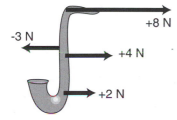

In this example, positive signs have been used consistently for forces to the right, and negative signs for forces to the left. The numerical value of a force carries no information about the place on the saxophone where the force is applied.

dimensional coordinate system with its x axis parallel the direction of motion. Forces that point along the positive x axis are positive, and forces in the opposite direction are negative. Forces that are not directly along the x axis cannot be immediately incorporated into this scheme, but that's OK, because we're avoiding those cases for now.

Discussion questions

In chapter 0, I defined 1 N as the force that would accelerate a 1-kg mass from rest to 1 m/s in 1 s. Anticipating the following section, you might guess that 2 N could be defined as the force that would accelerate the same mass to twice the speed, or twice the mass to the same speed. Is there an easier way to define 2 N based on the definition of 1 N?

4.2   Newton's First Law

We are now prepared to make a more powerful restatement of the principle of inertia.

Newton's First Law
If the total force on an object is zero, its center of mass continues in the same state of motion.

In other words, an object initially at rest is predicted to remain at rest if the total force on it is zero, and an object in motion remains in motion with the same velocity in the same direction. The converse of Newton's first law is also true: if we observe an object moving with constant velocity along a straight line, then the total force on it must be zero.

In a future physics course or in another textbook, you may encounter the term *net force*, which is simply a synonym for *total force*.

What happens if the total force on an object is not zero? It accelerates. Numerical prediction of the resulting acceleration is the topic of Newton's second law, which we'll discuss in the following section.

This is the first of Newton's three laws of motion. It is not important to memorize which of Newton's three laws are numbers one, two, and three. If a future physics teacher asks you something like, "Which of Newton's laws are you thinking of," a perfectly acceptable answer is "The one about constant velocity when there's zero total force." The concepts are more important than any specific formulation of them. Newton wrote in Latin, and I am not aware of any modern textbook that uses a verbatim translation of his statement of the laws of motion. Clear writing was not in vogue in Newton's day, and he formulated his three laws in terms of a concept now called momentum, only later relating it to the concept of force. Nearly all modern texts, including this one, start with force and do momentum later.

Example: an elevator
Question: An elevator has a weight of 5000 N. Compare the forces that the cable must exert to raise it at constant velocity, lower it at constant velocity, and just keep it hanging.
Answer: In all three cases the cable must pull up with a force of exactly 5000 N. Most people think you'd need at least a little more than 5000 N to make it go up, and a little less than 5000 N to let it down, but that's incorrect. Extra force from the cable is

only necessary for speeding the car up when it starts going up or slowing it down when it finishes going down. Decreased force is needed to speed the car up when it gets going down and to slow it down when it finishes going up. But when the elevator is cruising at constant velocity, Newton's first law says that you just need to cancel the force of the earth's gravity.

To many students, the statement in the example that the cable's upward force "cancels" the earth's downward gravitational force implies that there has been a contest, and the cable's force has won, vanquishing the earth's gravitational force and making it disappear. That is incorrect. Both forces continue to exist, but because they add up numerically to zero, the elevator has no center-of-mass acceleration. We know that both forces continue to exist because they both have side-effects other than their effects on the car's center-of-mass motion. The force acting between the cable and the car continues to produce tension in the cable and keep the cable taut. The earth's gravitational force continues to keep the passengers (whom we are considering as part of the elevator-object) stuck to the floor and to produce internal stresses in the walls of the car, which must hold up the floor.

Example: terminal velocity for falling objects
Question: An object like a feather that is not dense or streamlined does not fall with constant acceleration, because air resistance is nonnegligible. In fact, its acceleration tapers off to nearly zero within a fraction of a second, and the feather finishes dropping at constant speed (known as its terminal velocity). Why does this happen?

Answer: Newton's first law tells us that the total force on the feather must have been reduced to nearly zero after a short time. There are two forces acting on the feather: a downward gravitational force from the planet earth, and an upward frictional force from the air. As the feather speeds up, the air friction becomes stronger and stronger, and eventually it cancels out the earth's gravitational force, so the feather just continues with constant velocity without speeding up any more.

The situation for a skydiver is exactly analogous. It's just that the skydiver experiences perhaps a million times more gravitational force than the feather, and it is not until she is falling very fast that the force of air friction becomes as strong as the gravitational force. It takes her several seconds to reach terminal velocity, which is on the order of a hundred miles per hour.

More general combinations of forces

It is too constraining to restrict our attention to cases where all the forces lie along the line of the center of mass's motion. For one thing, we can't analyze any case of horizontal motion, since any object on earth will be subject to a vertical gravitational force! For instance, when you are driving your car down a straight road, there are both horizontal forces and vertical forces. However, the vertical forces have no effect on the center of mass motion, because the road's upward force simply counteracts the earth's downward gravitational force and keeps the car from sinking into the ground.

Later in the book we'll deal with the most general case of many forces acting on an object at any angles, using the mathematical technique of vector addition, but the following slight generalization of Newton's first law allows us to analyze a great many cases of interest:

Suppose that an object has two sets of forces acting on it, one set along the line of the object's initial motion and another set perpendicular to the first set. If both sets of forces cancel, then the object's center of mass continues in the same state of motion.

Example: a car crash
Question: If you drive your car into a brick wall, what is the mysterious force that slams your face into the steering wheel?
Answer: Your surgeon has taken physics, so she is not going to believe your claim that a mysterious force is to blame. She knows that your face was just following Newton's first law. Immediately after your car hit the wall, the only forces acting on your head were the same canceling-out forces that had existed previously: the earth's downward gravitational force and the upward force from your neck. There were no forward or backward forces on your head, but the car did experience a backward force from the wall, so the car slowed down and your face caught up.

Example: a passenger riding the subway
Question: Describe the forces acting on a person standing in a subway train that is cruising at constant velocity.
Answer: No force is necessary to keep the person moving relative to the ground. He will not be swept to the back of the train if the floor is slippery. There are two vertical forces on him, the earth's downward gravitational force and the floor's upward force, which cancel. There are no horizontal forces on him at all, so of course the total horizontal force is zero.

Example: forces on a sailboat
Question: If a sailboat is cruising at constant velocity with the wind coming from directly behind it, what must be true about the forces acting on it?
Answer: The forces acting on the boat must be canceling each other out. The boat is not sinking or leaping into the air, so evidently the vertical forces are canceling out. The vertical forces are the downward gravitational force exerted by the planet earth and an upward force from the water.

The air is making a forward force on the sail, and if the boat is not accelerating horizontally then the water's backward

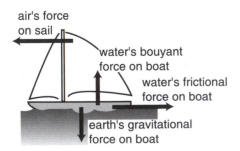

air's force on sail

water's bouyant force on boat

water's frictional force on boat

earth's gravitational force on boat

frictional force must be canceling it out.

Contrary to Aristotle, more force is not needed in order to maintain a higher speed. Zero total force is always needed to maintain constant velocity. Consider the following made-up numbers:

	boat moving at a low, constant velocity	boat moving at a high, constant velocity
forward force of the wind on the sail......	10,000 N	20,000 N
backward force of the water on the hull........................	-10,000 N	-20,000 N
total force on the boat......................	0 N	0 N

The faster boat still has zero total force on it. The forward force on it is greater, and the backward force smaller (more negative), but that's irrelevant because Newton's first law has to do with the total force, not the individual forces.

This example is quite analogous to the one about terminal velocity of falling objects, since there is a frictional force that increases with speed. After casting off from the dock and raising the sail, the boat will accelerate briefly, and then reach its terminal velocity, at which the water's frictional force has become as great as the wind's force on the sail.

Discussion questions

A. Newton said that objects continue moving if no forces are acting on them, but his predecessor Aristotle said that a force was necessary to keep an object moving. Why does Aristotle's theory seem more plausible, even though we now believe it to be wrong? What insight was Aristotle missing about the reason why things seem to slow down naturally?

B. In the first figure, what would have to be true about the saxophone's initial motion if the forces shown were to result in continued one-dimensional motion?

C. The second figure requires an ever further generalization of the preceding discussion. After studying the forces, what does your physical intuition tell you will happen? Can you state in words how to generalize the conditions for one-dimensional motion to include situations like this one?

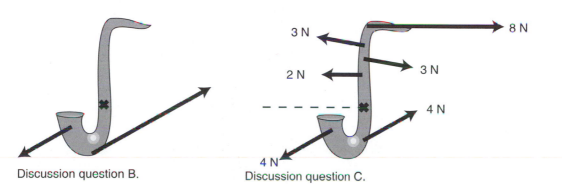

Discussion question B.

Discussion question C.

4.3   Newton's Second Law

What about cases where the total force on an object is not zero, so that Newton's first law doesn't apply? The object will have an acceleration. The way we've defined positive and negative signs of force and acceleration guarantees that positive forces produce positive accelerations, and likewise for negative values. How much acceleration will it have? It will clearly depend on both the object's mass and on the amount of force.

Experiments with any particular object show that its acceleration is directly proportional to the total force applied to it. This may seem wrong, since we know of many cases where small amounts of force fail to move an object at all, and larger forces get it going. This apparent failure of proportionality actually results from forgetting that there is a frictional force in addition to the force we apply to move the object. The object's acceleration is exactly proportional to the total force on it, not to any individual force on it. In the absence of friction, even a very tiny force can slowly change the velocity of a very massive object.

Experiments also show that the acceleration is inversely proportional to the object's mass, and combining these two proportionalities gives the following way of predicting the acceleration of any object:

Newton's Second Law

$$a = F_{total}/m \quad,$$

where

m is an object's mass

F_{total} is the sum of the forces acting on it, and

a is the acceleration of the object's center of mass.

We are presently restricted to the case where the forces of interest are parallel to the direction of motion.

Example: an accelerating bus

Question: A VW bus with a mass of 2000 kg accelerates from 0 to 25 m/s (freeway speed) in 34 s. Assuming the acceleration is constant, what is the total force on the bus?

Solution: We solve Newton's second law for $F_{total}=ma$, and substitute $\Delta v/\Delta t$ for a, giving

$$\begin{aligned} F_{total} &= m\Delta v/\Delta t \\ &= (2000\ \text{kg})(25\ \text{m/s} - 0\ \text{m/s})/(34\ \text{s}) \\ &= 1.5\ \text{kN} \quad. \end{aligned}$$

A generalization

As with the first law, the second law can be easily generalized to include a much larger class of interesting situations:

Suppose an object is being acted on by two sets of forces, one set lying along the object's initial direction of motion and another set acting along a perpendicular line. If the forces perpendicular to the initial direction of motion cancel out, then the object accelerates along its original line of motion according to $a=F_{total}/m$.

The relationship between mass and weight

Mass is different from weight, but they're related. An apple's mass tells

A simple double-pan balance works by comparing the weight forces exerted by the earth on the contents of the two pans. Since the two pans are at almost the same location on the earth's surface, the value of g is essentially the same for each one, and equality of weight therefore also implies equality of mass.

us how hard it is to change its motion. Its weight measures the strength of the gravitational attraction between the apple and the planet earth. The apple's weight is less on the moon, but its mass is the same. Astronauts assembling the International Space Station in zero gravity cannot just pitch massive modules back and forth with their bare hands; the modules are weightless, but not massless.

We have already seen the experimental evidence that when weight (the force of the earth's gravity) is the only force acting on an object, its acceleration equals the constant g, and g depends on where you are on the surface of the earth, but not on the mass of the object. Applying Newton's second law then allows us to calculate the magnitude of the gravitational force on any object in terms of its mass:

$$|F_W| = mg \ .$$

(The equation only gives the magnitude, i.e. the absolute value, of F_W, because we're defining g as a positive number, so it equals the absolute value of a falling object's acceleration.)

Example: calculating terminal velocity

Question: Experiments show that the force of air friction on a falling object such as a skydiver or a feather can be approximated fairly well with the equation $|F_{air}|=c\rho Av^2$, where c is a constant, ρ is the density of the air, A is the cross-sectional area of the object as seen from below, and v is the object's velocity. Predict the object's terminal velocity, i.e. the final velocity it reaches after a long time.

Solution: As the object accelerates, its greater v causes the upward force of the air to increase until finally the gravitational force and the force of air friction cancel out, after which the object continues at constant velocity. We choose a coordinate system in which positive is up, so that the gravitational force is negative and the force of air friction is positive. We want to find the velocity at which

$$F_{air} + F_W \ = \ 0 \ , \text{i.e.}$$
$$c\rho Av^2 - mg = \ 0 \ .$$

Solving for v gives

$$v_{terminal} = \sqrt{\frac{mg}{c\rho A}}$$

Self-Check

It is important to get into the habit of interpreting equations. These two self-check questions may be difficult for you, but eventually you will get used to this kind of reasoning.

(a) Interpret the equation $v_{terminal} = \sqrt{mg/c\rho A}$ in the case of $\rho=0$.

(b) How would the terminal velocity of a 4-cm steel ball compare to that of a 1-cm ball?

 (a) The case of $\rho=0$ represents an object falling in a vacuum, i.e. there is no density of air. The terminal velocity would be infinite. Physically, we know that an object falling in a vacuum would never stop speeding up, since there would be no force of air friction to cancel the force of gravity. (b) The 4-cm ball would have a mass that was greater by a factor of 4x4x4, but its cross-sectional area would be greater by a factor of 4x4. Its terminal velocity would be greater by a factor of $\sqrt{4^3 / 4^2} = 2$.

A. Show that the Newton can be reexpressed in terms of the three basic mks units as the combination kg·m/s^2.

B. What is wrong with the following statements?

1. "g is the force of gravity."

2. "Mass is a measure of how much space something takes up."

C. Criticize the following incorrect statement:

"If an object is at rest and the total force on it is zero, it stays at rest. There can also be cases where an object is moving and keeps on moving without having any total force on it, but that can only happen when there's no friction, like in outer space."

D. The table on the left gives laser timing data for Ben Johnson's 100 m dash at the 1987 World Championship in Rome. (His world record was later revoked because he tested positive for steroids.) How does the total force on him change over the duration of the race?

x (m)	t (s)
10	1.84
20	2.86
30	3.80
40	4.67
50	5.53
60	6.38
70	7.23
80	8.10
90	8.96
100	9.83

Discussion question D.

4.4 What Force Is Not

Violin teachers have to endure their beginning students' screeching. A frown appears on the woodwind teacher's face as she watches her student take a breath with an expansion of his ribcage but none in his belly. What makes physics teachers cringe is their students' verbal statements about forces. Below I have listed several dicta about what force is not.

Force is not a property of one object.

A great many of students' incorrect descriptions of forces could be cured by keeping in mind that a force is an interaction of two objects, not a property of one object.

> *Incorrect statement*: "That magnet has a lot of force."
> ✗ If the magnet is one millimeter away from a steel ball bearing, they may exert a very strong attraction on each other, but if they were a meter apart, the force would be virtually undetectable. The magnet's strength can be rated using certain electrical units (ampere-meters^2), but not in units of force.

Force is not a measure of an object's motion.

If force is not a property of a single object, then it cannot be used as a measure of the object's motion.

> *Incorrect statement*: "The freight train rumbled down the tracks with awesome force."
> ✗ Force is not a measure of motion. If the freight train collides with a stalled cement truck, then some awesome forces will occur, but if it hits a fly the force will be small.

Force is not energy.

There are two main approaches to understanding the motion of objects, one based on force and one on a different concept, called energy. The SI unit of energy is the Joule, but you are probably more familiar with the calorie, used for measuring food's energy, and the kilowatt-hour, the unit the electric company uses for billing you. Physics students' previous familiarity with calories and kilowatt-hours is matched by their universal unfamiliarity with measuring forces in units of Newtons, but the precise operational definitions of the energy concepts are more complex than those of the force concepts, and textbooks, including this one, almost universally place the force description of physics before the energy description. During the long period after the introduction of force and before the careful definition of energy, students are therefore vulnerable to situations in which, without realizing it, they are imputing the properties of energy to phenomena of force.

> *Incorrect statement:* "How can my chair be making an upward force on my rear end? It has no power!"
> ✗ Power is a concept related to energy, e.g. 100-watt lightbulb uses up 100 joules per second of energy. When you sit in a chair, no energy is used up, so forces can exist between you and the chair without any need for a source of power.

Force is not stored or used up.

Because energy can be stored and used up, people think force also can be stored or used up.

> *Incorrect statement:* "If you don't fill up your tank with gas, you'll run out of force."
> ✗ Energy is what you'll run out of, not force.

Forces need not be exerted by living things or machines.

Transforming energy from one form into another usually requires some kind of living or mechanical mechanism. The concept is not applicable to forces, which are an interaction between objects, not a thing to be transferred or transformed.

> *Incorrect statement:* "How can a wooden bench be making an upward force on my rear end? It doesn't have any springs or anything inside it."
> ✗ No springs or other internal mechanisms are required. If the bench didn't make any force on you, you would obey Newton's second law and fall through it. Evidently it does make a force on you!

A force is the direct cause of a change in motion.

I can click a remote control to make my garage door change from being at rest to being in motion. My finger's force on the button, however, was not the force that acted on the door. When we speak of a force on an object in physics, we are talking about a force that acts directly. Similarly, when you pull a reluctant dog along by its leash, the leash and the dog are making forces on each other, not your hand and the dog. The dog is not even touching your hand.

Self-Check

Which of the following things can be correctly described in terms of force?
(a) A nuclear submarine is charging ahead at full steam.
(b) A nuclear submarine's propellers spin in the water.
(c) A nuclear submarine needs to refuel its reactor periodically.

Discussion questions

A. Criticize the following incorrect statement: "If you shove a book across a table, friction takes away more and more of its force, until finally it stops."
B. You hit a tennis ball against a wall. Explain any and all incorrect ideas in the following description of the physics involved: "The ball gets some force from you when you hit it, and when it hits the wall, it loses part of that force, so it doesn't bounce back as fast. The muscles in your arm are the only things that a force can come from."

(a) This is motion, not force. (b) This is a description of how the sub is able to get the water to produce a forward force on it. (c) The sub runs out of energy, not force.

4.5 Inertial and Noninertial Frames of Reference

One day, you're driving down the street in your pickup truck, on your way to deliver a bowling ball. The ball is in the back of the truck, enjoying its little jaunt and taking in the fresh air and sunshine. Then you have to slow down because a stop sign is coming up. As you brake, you glance in your rearview mirror, and see your trusty companion accelerating toward you. Did some mysterious force push it forward? No, it only seems that way because you and the car are slowing down. The ball is faithfully obeying Newton's first law, and as it continues at constant velocity it gets ahead relative to the slowing truck. No forces are acting on it (other than the same canceling-out vertical forces that were always acting on it). The ball only appeared to violate Newton's first law because there was something wrong with your frame of reference, which was based on the truck.

How, then, are we to tell in which frames of reference Newton's laws are valid? It's no good to say that we should avoid moving frames of reference, because there is no such thing as absolute rest or absolute motion. All frames can be considered as being either at rest or in motion. According to

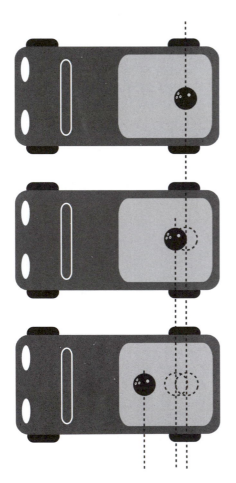

(a) In a frame of reference that moves with the truck, the bowling ball appears to violate Newton's first law by accelerating despite having no horizontal forces on it.

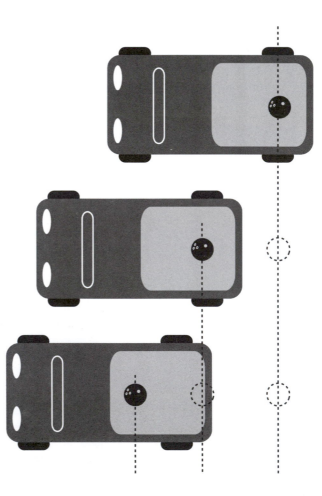

(b) In an inertial frame of reference, which the surface of the earth approximately is, the bowling ball obeys Newton's first law. It moves equal distances in equal time intervals, i.e. maintains constant velocity. In this frame of reference, it is the truck that appears to have a change in velocity, which makes sense, since the road is making a horizontal force on it.

an observer in India, the strip mall that constituted the frame of reference in panel (b) of the figure was moving along with the earth's rotation at hundreds of miles per hour.

The reason why Newton's laws fail in the truck's frame of reference is not because the truck is *moving* but because it is *accelerating*. (Recall that physicists use the word to refer either to speeding up or slowing down.) Newton's laws were working just fine in the moving truck's frame of reference as long as the truck was moving at constant velocity. It was only when its speed changed that there was a problem. How, then, are we to tell which frames are accelerating and which are not? What if you claim that your truck is not accelerating, and the sidewalk, the asphalt, and the Burger King are accelerating? The way to settle such a dispute is to examine the motion of some object, such as the bowling ball, which we know has zero total force on it. Any frame of reference in which the ball appears to obey Newton's first law is then a valid frame of reference, and to an observer in that frame, Mr. Newton assures us that all the other objects in the universe will obey his laws of motion, not just the ball.

Valid frames of reference, in which Newton's laws are obeyed, are called *inertial frames of reference.* Frames of reference that are not inertial are called noninertial frames. In those frames, objects violate the principle of inertia and Newton's first law. While the truck was moving at constant velocity, both it and the sidewalk were valid inertial frames. The truck became an invalid frame of reference when it began changing its velocity.

You usually assume the ground under your feet is a perfectly inertial frame of reference, and we made that assumption above. It isn't perfectly inertial, however. Its motion through space is quite complicated, being composed of a part due to the earth's daily rotation around its own axis, the monthly wobble of the planet caused by the moon's gravity, and the rotation of the earth around the sun. Since the accelerations involved are numerically small, the earth is approximately a valid inertial frame.

Noninertial frames are avoided whenever possible, and we will seldom, if ever, have occasion to use them in this course. Sometimes, however, a noninertial frame can be convenient. Naval gunners, for instance, get all their data from radars, human eyeballs, and other detection systems that are moving along with the earth's surface. Since their guns have ranges of many miles, the small discrepancies between their shells' actual accelerations and the accelerations predicted by Newton's second law can have effects that accumulate and become significant. In order to kill the people they want to kill, they have to add small corrections onto the equation $a = F_{total}/m$. Doing their calculations in an inertial frame would allow them to use the usual form of Newton's second law, but they would have to convert all their data into a different frame of reference, which would require cumbersome calculations.

Discussion question

If an object has a linear x-t graph in a certain inertial frame, what is the effect on the graph if we change to a coordinate system with a different origin? What is the effect if we keep the same origin but reverse the positive direction of the x axis? How about an inertial frame moving alongside the object? What if we describe the object's motion in a noninertial frame?

Summary

Selected Vocabulary

weight .... the force of gravity on an object, equal to mg

inertial frame .... a frame of reference that is not accelerating, one in which Newton's first law is true

noninertial frame .... an accelerating frame of reference, in which Newton's first law is violated

Terminology Used in Some Other Books

net force .... another way of saying "total force"

Notation

F_W .... the weight force

Summary

Newton's first law of motion states that if all the forces on an object cancel each other out, then the object continues in the same state of motion. This is essentially a more refined version of Galileo's principle of inertia, which did not refer to a numerical scale of force.

Newton's second law of motion allows the prediction of an object's acceleration given its mass and the total force on it, $a = F_{total}/m$. This is only the one-dimensional version of the law; the full-three dimensional treatment will come in chapter 8, Vectors. Without the vector techniques, we can still say that the situation remains unchanged by including an additional set of vectors that cancel among themselves, even if they are not in the direction of motion.

Newton's laws of motion are only true in frames of reference that are not accelerating, known as inertial frames.

Homework Problems

1. An object is observed to be moving at constant speed in a certain direction. Can you conclude that no forces are acting on it? Explain. [Based on a problem by Serway and Faughn.]

2. A car is normally capable of an acceleration of 3 m/s². If it is towing a trailer with half as much mass as the car itself, what acceleration can it achieve? [Based on a problem from PSSC Physics.]

3. (a✓) Let T be the maximum tension that the elevator's cable can withstand without breaking, i.e. the maximum force it can exert. If the motor is programmed to give the car an acceleration a, what is the maximum mass that the car can have, including passengers, if the cable is not to break? (b) Interpret the equation you derived in the special cases of $a=0$ and of a downward acceleration of magnitude g.

4✓. A helicopter of mass m is taking off vertically. The only forces acting on it are the earth's gravitational force and the force, F_{air}, of the air pushing up on the propeller blades. (a) If the helicopter lifts off at $t=0$, what is its vertical speed at time t? (b) Plug numbers into your equation from part a, using $m=2300$ kg, $F_{air}=27000$ N, and $t=4.0$ s.

5★. In the 1964 Olympics in Tokyo, the best men's high jump was 2.18 m. Four years later in Mexico City, the gold medal in the same event was for a jump of 2.24 m. Because of Mexico City's altitude (2400 m), the acceleration of gravity there is lower than that in Tokyo by about 0.01 m/s². Suppose a high-jumper has a mass of 72 kg.

(a) Compare his mass and weight in the two locations.

(b✓) Assume that he is able to jump with the same initial vertical velocity in both locations, and that all other conditions are the same except for gravity. How much higher should he be able to jump in Mexico City?

(Actually, the reason for the big change between '64 and '68 was the introduction of the "Fosbury flop.")

6∫✓. A blimp is initially at rest, hovering, when at $t=0$ the pilot turns on the motor of the propeller. The motor cannot instantly get the propeller going, but the propeller speeds up steadily. The steadily increasing force between the air and the propeller is given by the equation $F=kt$, where k is a constant. If the mass of the blimp is m, find its position as a function of time. (Assume that during the period of time you're dealing with, the blimp is not yet moving fast enough to cause a significant backward force due to air resistance.)

Problem 6.

7 S. A car is accelerating forward along a straight road. If the force of the road on the car's wheels, pushing it forward, is a constant 3.0 kN, and the car's mass is 1000 kg, then how long will the car take to go from 20 m/s to 50 m/s?

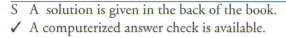

S A solution is given in the back of the book.   ★ A difficult problem.

✓ A computerized answer check is available.   ∫ A problem that requires calculus.

8. Some garden shears are like a pair of scissors: one sharp blade slices past another. In the "anvil" type, however, a sharp blade presses against a flat one rather than going past it. A gardening book says that for people who are not very physically strong, the anvil type can make it easier to cut tough branches, because it concentrates the force on one side. Evaluate this claim based on Newton's laws. [Hint: Consider the forces acting on the branch, and the motion of the branch.]

9. A uranium atom deep in the earth spits out an alpha particle. An alpha particle is a fragment of an atom. This alpha particle has initial speed v, and travels a distance d before stopping in the earth. (a) Find the force, F, that acted on the particle, in terms of v, d, and its mass, m. Don't plug in any numbers yet. Assume that the force was constant. (b) Show that your answer has the right units. (c) Discuss how your answer to part a depends on all three variables, and show that it makes sense. (d) Evaluate your result for $m=6.7 \times 10^{-27}$ kg, $v=2.0 \times 10^4$ km/s, and $d=0.71$ mm.

Motion in Three Dimensions

Photo by Clarence White, ca. 1903.

6 Newton's Laws in Three Dimensions

6.1 Forces Have No Perpendicular Effects

Suppose you could shoot a rifle and arrange for a second bullet to be dropped from the same height at the exact moment when the first left the barrel. Which would hit the ground first? Nearly everyone expects that the dropped bullet will reach the dirt first, and Aristotle would have agreed. Aristotle would have described it like this. The shot bullet receives some forced motion from the gun. It travels forward for a split second, slowing

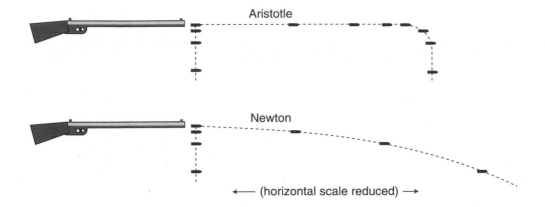

← (horizontal scale reduced) →

down rapidly because there is no longer any force to make it continue in motion. Once it is done with its forced motion, it changes to natural motion, i.e. falling straight down. While the shot bullet is slowing down, the dropped bullet gets on with the business of falling, so according to Aristotle it will hit the ground first.

Luckily, nature isn't as complicated as Aristotle thought! To convince yourself that Aristotle's ideas were wrong and needlessly complex, stand up now and try this experiment. Take your keys out of your pocket, and begin walking briskly forward. Without speeding up or slowing down, release your keys and let them fall while you continue walking at the same pace.

You have found that your keys hit the ground right next to your feet. Their horizontal motion never slowed down at all, and the whole time they were dropping, they were right next to you. The horizontal motion and the vertical motion happen at the same time, and they are independent of each other. Your experiment proves that the horizontal motion is unaffected by the vertical motion, but it's also true that the vertical motion is not changed in any way by the horizontal motion. The keys take exactly the same amount of time to get to the ground as they would have if you simply dropped them, and the same is true of the bullets: both bullets hit the ground simultaneously.

These have been our first examples of motion in more than one dimension, and they illustrate the most important new idea that is required to understand the three-dimensional generalization of Newtonian physics:

> **Forces have no perpendicular effects.**
> When a force acts on an object, it has no effect on the part of the object's motion that is perpendicular to the force.

In the examples above, the vertical force of gravity had no effect on the horizontal motions of the objects. These were examples of projectile motion, which interested people like Galileo because of its military applications. The principle is more general than that, however. For instance, if a rolling ball is initially heading straight for a wall, but a steady wind begins blowing from the side, the ball does not take any longer to get to the wall. In the case of projectile motion, the force involved is gravity, so we can say more specifically that the vertical acceleration is 9.8 m/s², regardless of the horizontal motion.

Relationship to relative motion

These concepts are directly related to the idea that motion is relative. Galileo's opponents argued that the earth could not possibly be rotating as he claimed, because then if you jumped straight up in the air you wouldn't be able to come down in the same place. Their argument was based on their incorrect Aristotelian assumption that once the force of gravity began to act on you and bring you back down, your horizontal motion would stop. In the correct Newtonian theory, the earth's downward gravitational force is acting before, during, and after your jump, but has no effect on your motion in the perpendicular (horizontal) direction.

If Aristotle had been correct, then we would have a handy way to determine absolute motion and absolute rest: jump straight up in the air, and if you land back where you started, the surface from which you jumped must have been in a state of rest. In reality, this test gives the same result as long as the surface under you is an inertial frame. If you try this in a jet plane, you land back on the same spot on the deck from which you started, regardless of whether the plane is flying at 500 miles per hour or parked on the runway. The method would in fact only be good for detecting whether the plane was accelerating.

Discussion Questions

A. The following is an incorrect explanation of a fact about target shooting: "Shooting a high-powered rifle with a high muzzle velocity is different from shooting a less powerful gun. With a less powerful gun, you have to aim quite a bit above your target, but with a more powerful one you don't have to aim so high because the bullet doesn't drop as fast."
What is the correct explanation?

B. You have thrown a rock, and it is flying through the air in an arc. If the earth's gravitational force on it is always straight down, why doesn't it just go straight down once it leaves your hand?
C. Consider the example of the bullet that is dropped at the same moment another bullet is fired from a gun. What would the motion of the two bullets look like to a jet pilot flying alongside in the same direction as the shot bullet and at the same horizontal speed?

6.2 Coordinates and Components

'Cause we're all
Bold as love,
Just ask the axis.

-Jimi Hendrix

How do we convert these ideas into mathematics? The figure below shows a good way of connecting the intuitive ideas to the numbers. In one dimension, we impose a number line with an x coordinate on a certain stretch of space. In two dimensions, we imagine a grid of squares which we label with x and y values.

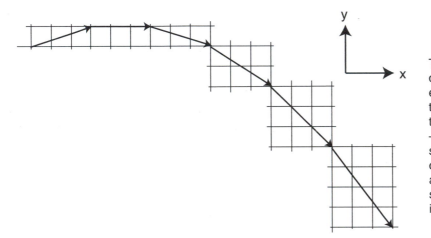

This object experiences a force that pulls it down toward the bottom of the page. In each equal time interval, it moves three units to the right. At the same time, its vertical motion is making a simple pattern of +1, 0, −1, −2, −3, −4, ... units. Its motion can be described by an x coordinate that has zero acceleration and a y coordinate with constant acceleration. The arrows labeled x and y serve to explain that we are defining increasing x to the right and increasing y as upward.

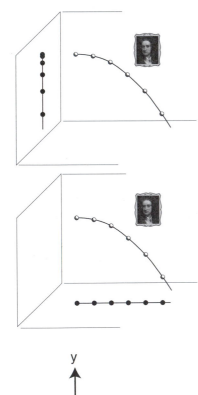

But of course motion doesn't really occur in a series of discrete hops like in chess or checkers. The figure on the left shows a way of conceptualizing the smooth variation of the x and y coordinates. The ball's shadow on the wall moves along a line, and we describe its position with a single coordinate, y, its height above the floor. The wall shadow has a constant acceleration of −9.8 m/s². A shadow on the floor, made by a second light source, also moves along a line, and we describe its motion with an x coordinate, measured from the wall.

The velocity of the floor shadow is referred to as the *x component* of the velocity, written v_x. Similarly we can notate the acceleration of the floor shadow as a_x. Since v_x is constant, a_x is zero.

Similarly, the velocity of the wall shadow is called v_y, its acceleration a_y. This example has $a_y = -9.8$ m/s².

Because the earth's gravitational force on the ball is acting along the y axis, we say that the force has a negative y component, F_y, but $F_x = F_z = 0$.

The general idea is that we imagine two observers, each of whom perceives the entire universe as if it was flattened down to a single line. The y-observer, for instance, perceives y, v_y, and a_y, and will infer that there is a force, F_y, acting downward on the ball. That is, a y component means the aspect of a physical phenomenon, such as velocity, acceleration, or force, that is observable to someone who can only see motion along the y axis.

All of this can easily be generalized to three dimensions. In the example above, there could be a z-observer who only sees motion toward or away from the back wall of the room.

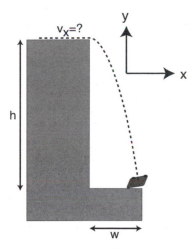

Example: a car going over a cliff

Question: The police find a car at a distance w=20 m from the base of a cliff of height h=100 m. How fast was the car going when it went over the edge? Solve the problem symbolically first, then plug in the numbers.

Solution: Let's choose y pointing up and x pointing away from the cliff. The car's vertical motion was independent of its horizontal motion, so we know it had a constant vertical acceleration of a=-g=-9.8 m/s². The time it spent in the air is therefore related to the vertical distance it fell by the constant-acceleration equation

$$\Delta y = \frac{1}{2} a_y \Delta t^2 \quad ,$$

or

$$-h = \frac{1}{2}(-g)\Delta t^2 \quad .$$

Solving for Δt gives

$$\Delta t = \sqrt{\frac{2h}{g}} \quad .$$

Since the vertical force had no effect on the car's horizontal motion, it had a_x=0, i.e. constant horizontal velocity. We can apply the constant-velocity equation

$$v_x = \frac{\Delta x}{\Delta t} \quad ,$$

i.e.

$$v_x = \frac{w}{\Delta t} \quad .$$

We now substitute for Δt to find

$$v_x = w / \sqrt{\frac{2h}{g}} \quad ,$$

which simplifies to

$$v_x = w\sqrt{\frac{g}{2h}} \quad .$$

Plugging in numbers, we find that the car's speed when it went over the edge was 4 m/s, or about 10 mi/hr.

Projectiles move along parabolas

What type of mathematical curve does a projectile follow through space? To find out, we must relate x to y, eliminating t. The reasoning is very similar to that used in the example above. Arbitrarily choosing $x=y=t=0$ to be at the top of the arc, we conveniently have $x=\Delta x$, $y=\Delta y$, and $t=\Delta t$, so

$$y = -\frac{1}{2}a_y t^2$$

$$x = v_x t$$

We solve the second equation for $t=x/v_x$ and eliminate t in the first equation:

$$y = -\frac{1}{2}a_y\left(\frac{x}{v_x}\right)^2 \quad .$$

Since everything in this equation is a constant except for x and y, we conclude that y is proportional to the square of x. As you may or may not recall from a math class, $y \propto x^2$ describes a parabola.

A parabola can be defined as the shape made by cutting a cone parallel to its side. A parabola is also the graph of an equation of the form $y \propto x^2$.

Each water droplet follows a parabola. The faster drops' parabolas are bigger.

Discussion Question

A. At the beginning of this section I represented the motion of a projectile on graph paper, breaking its motion into equal time intervals. Suppose instead that there is no force on the object at all. It obeys Newton's first law and continues without changing its state of motion. What would the corresponding graph-paper diagram look like? If the time interval represented by each arrow was 1 second, how would you relate the graph-paper diagram to the velocity components v_x and v_y?

B. Make up several different coordinate systems oriented in different ways, and describe the a_x and a_y of a falling object in each one.

6.3 Newton's Laws in Three Dimensions

It is now fairly straightforward to extend Newton's laws to three dimensions:

Newton's First Law

If all three components of the total force on an object are zero, then it will continue in the same state of motion.

Newton's Second Law

An object's acceleration components are predicted by the equations

$$a_x = F_{x,\text{total}}/m \ ,$$

$$a_y = F_{y,\text{total}}/m \ , \text{ and}$$

$$a_z = F_{z,\text{total}}/m \ .$$

Newton's Third Law

If two objects A and B interact via forces, then the components of their forces on each other are equal and opposite:

$$F_{\text{A on B},x} = -F_{\text{B on A},x} \ ,$$

$$F_{\text{A on B},y} = -F_{\text{B on A},y} \ , \text{ and}$$

$$F_{\text{A on B},z} = -F_{\text{B on A},z} \ .$$

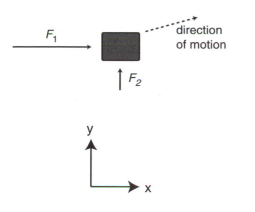

direction of motion

Example: forces in perpendicular directions on the same object
Question: An object is initially at rest. Two constant forces begin acting on it, and continue acting on it for a while. As suggested by the two arrows, the forces are perpendicular, and the rightward force is stronger. What happens?

Answer: Aristotle believed, and many students still do, that only one force can "give orders" to an object at one time. They therefore think that the object will begin speeding up and moving in the direction of the stronger force. In fact the object will move along a diagonal. In the example shown in the figure, the object will respond to the large rightward force with a large acceleration component to the right, and the small upward force will give it a small acceleration component upward. The stronger force does not overwhelm the weaker force, or have any effect on the upward motion at all. The force components simply add together:

$$F_{x,total} = F_{1,x} + \overbrace{F_{2,x}}^{=0}$$

$$F_{y,total} = \overbrace{F_{1,y}}^{=0} + F_{2,y}$$

Discussion Question

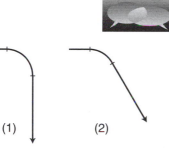

The figure shows two trajectories, made by splicing together lines and circular arcs, which are unphysical for an object that is only being acted on by gravity. Prove that they are impossible based on Newton's laws.

(1)   (2)

Summary

Selected Vocabulary

        component .... the part of a velocity, acceleration, or force that would be perceptible to an observer who could only see the universe projected along a certain one-dimensional axis

        parabola .... the mathematical curve whose graph has y proportional to x^2

Notation

        x, y, z .... an object's positions along the x, y, and z axes

        v_x, v_y, v_z .... the x, y, and z components of an object's velocity; the rates of change of the object's x, y, and z coordinates

        a_x, a_y, a_z .... the x, y, and z components of an object's acceleration; the rates of change of v_x, v_y, and v_z

Summary

A force does not produce any effect on the motion of an object in a perpendicular direction. The most important application of this principle is that the horizontal motion of a projectile has zero acceleration, while the vertical motion has an acceleration equal to g. That is, an object's horizontal and vertical motions are independent. The arc of a projectile is a parabola.

Motion in three dimensions is measured using three coordinates, x, y, and z. Each of these coordinates has its own corresponding velocity and acceleration. We say that the velocity and acceleration both have x, y, and z components

Newton's second law is readily extended to three dimensions by rewriting it as three equations predicting the three components of the acceleration,

$$a_x = F_{x,total}/m \ ,$$
$$a_y = F_{y,total}/m \ ,$$
$$a_z = F_{z,total}/m \ ,$$

and likewise for the first and third laws.

Homework Problems

1✓. (a) A ball is thrown straight up with velocity v. Find an equation for the height to which it rises.

(b) Generalize your equation for a ball thrown at an angle θ above horizontal, in which case its initial velocity components are $v_x = v \cos \theta$ and $v_y = v \sin \theta$.

2. At the Salinas Lettuce Festival Parade, Miss Lettuce of 1996 drops her bouquet while riding on a float. Compare the shape of its trajectory as seen by her to the shape seen by one of her admirers standing on the sidewalk.

3 . Two daredevils, Wendy and Bill, go over Niagara Falls. Wendy sits in an inner tube, and lets the 30 km/hr velocity of the river throw her out horizontally over the falls. Bill paddles a kayak, adding an extra 10 km/hr to his velocity. They go over the edge of the falls at the same moment, side by side. Ignore air friction. Explain your reasoning.

(a) Who hits the bottom first?

(b) What is the horizontal component of Wendy's velocity on impact?

(c) What is the horizontal component of Bill's velocity on impact?

(d) Who is going faster on impact?

4✓. A baseball pitcher throws a pitch clocked at v_x=73.3 mi/h. He throws horizontally. By what amount, d, does the ball drop by the time it reaches home plate, L=60.0 ft away? (a) First find a symbolic answer in terms of L, v_x, and g. (b) Plug in and find a numerical answer. Express your answer in units of ft. [Note: 1 ft=12 in, 1 mi=5280 ft, and 1 in=2.54 cm]

5 S. A cannon standing on a flat field fires a cannonball with a muzzle velocity v, at an angle θ above horizontal. The cannonball thus initially has velocity components $v_x = v \cos \theta$ and $v_y = v \sin \theta$.

(a) Show that the cannon's range (horizontal distance to where the cannonball falls) is given by the equation $R = \dfrac{2v^2 \sin \theta \cos \theta}{g}$.

(b) Interpret your equation in the cases of θ=0 and θ=90°.

S   A solution is given in the back of the book.   ★ A difficult problem.

✓ A computerized answer check is available.   ∫ A problem that requires calculus.

6 ∫. Assuming the result of the previous problem for the range of a projectile, $R = \dfrac{2v^2 \sin \theta \cos \theta}{g}$, show that the maximum range is for $\theta = 45°$.

7. Two cars go over the same bump in the road, Maria's Maserati at 25 miles per hour and Park's Porsche at 37. How many times greater is the vertical acceleration of the Porsche? Hint: Remember that acceleration depends both on how much the velocity changes and on how much time it takes to change.

Rockets work by pushing exhaust gases out the back. Newton's third law says that if the rocket exerts a backward force on the gases, the gases must make an equal forward force on the rocket. Rocket engines can function above the atmosphere, unlike propellers and jets, which work by pushing against the surrounding air.

5 Analysis of Forces

5.1 Newton's Third Law

Newton created the modern concept of force starting from his insight that all the effects that govern motion are interactions between two objects: unlike the Aristotelian theory, Newtonian physics has no phenomena in which an object changes its own motion.

Is one object always the "order-giver" and the other the "order-follower"? As an example, consider a batter hitting a baseball. The bat definitely exerts a large force on the ball, because the ball accelerates drastically. But if you have ever hit a baseball, you also know that the ball makes a force on the bat — often with painful results if your technique is as bad as mine!

How does the ball's force on the bat compare with the bat's force on the ball? The bat's acceleration is not as spectacular as the ball's, but maybe we shouldn't expect it to be, since the bat's mass is much greater. In fact, careful measurements of both objects' masses and accelerations would show that $m_{ball}a_{ball}$ is very nearly equal to $-m_{bat}a_{bat}$, which suggests that the ball's force on the bat is of the same magnitude as the bat's force on the ball, but in the opposite direction.

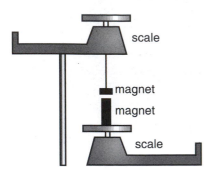

(a) Two magnets exert forces on each other.

(b) Two people's hands exert forces on each other.

The figures show two somewhat more practical laboratory experiments for investigating this issue accurately and without too much interference from extraneous forces.

In the first experiment, a large magnet and a small magnet are weighed separately, and then one magnet is hung from the pan of the top balance so that it is directly above the other magnet. There is an attraction between the two magnets, causing the reading on the top scale to increase and the reading on the bottom scale to decrease. The large magnet is more "powerful" in the sense that it can pick up a heavier paperclip from the same distance, so many people have a strong expectation that one scale's reading will change by a far different amount than the other. Instead, we find that the two changes are equal in magnitude but opposite in direction, so the upward force of the top magnet on the bottom magnet is of the same magnitude as the downward force of the bottom magnet on the top magnet.

In the second experiment, two people pull on two spring scales. Regardless of who tries to pull harder, the two forces as measured on the spring scales are equal. Interposing the two spring scales is necessary in order to measure the forces, but the outcome is not some artificial result of the scales' interactions with each other. If one person slaps another hard on the hand, the slapper's hand hurts just as much as the slappee's, and it doesn't matter if the recipient of the slap tries to be inactive. (Punching someone in the mouth causes just as much force on the fist as on the lips. It's just that the lips are more delicate. The forces are equal, but not the levels of pain and injury.)

Newton, after observing a series of results such as these, decided that there must be a fundamental law of nature at work:

> **Newton's Third Law**
> Forces occur in equal and opposite pairs: whenever object A exerts a force on object B, object B must also be exerting a force on object A. The two forces are equal in magnitude and opposite in direction.

In one-dimensional situations, we can use plus and minus signs to indicate the directions of forces, and Newton's third law can be written succinctly as
$$F_{A \text{ on } B} = -F_{B \text{ on } A}.$$

There is no cause and effect relationship between the two forces. There is no "original" force, and neither one is a response to the other. The pair of forces is a relationship, like marriage, not a back-and-forth process like a tennis match. Newton came up with the third law as a generalization about all the types of forces with which he was familiar, such as frictional and gravitational forces. When later physicists discovered a new type force, such as the force that holds atomic nuclei together, they had to check whether it obeyed Newton's third law. So far, no violation of the third law has ever been discovered, whereas the first and second laws were shown to have limitations by Einstein and the pioneers of atomic physics.

Newton's third law does not mean that forces always cancel out so that nothing can ever move. If these two figure skaters, initially at rest, push against each other, they will both move.

It doesn't make sense for the man to talk about the woman's money canceling out his bar tab, because there is no good reason to combine his debts and her assets. Similarly, it doesn't make sense to refer to the equal and opposite forces of Newton's third law as canceling. It only makes sense to add up forces that are acting on the *same* object, whereas two forces related to each other by Newton's third law are always acting on two *different* objects.

The English vocabulary for describing forces is unfortunately rooted in Aristotelianism, and often implies incorrectly that forces are one-way relationships. It is unfortunate that a half-truth such as "the table exerts an upward force on the book" is so easily expressed, while a more complete and correct description ends up sounding awkward or strange: "the table and the book interact via a force," or "the table and book participate in a force."

To students, it often sounds as though Newton's third law implies nothing could ever change its motion, since the two equal and opposite forces would always cancel. The two forces, however, are always on two different objects, so it doesn't make sense to add them in the first place — we only add forces that are acting on the same object. If two objects are interacting via a force and no other forces are involved, then *both* objects will accelerate — in opposite directions!

A mnemonic for using Newton's third law correctly

Mnemonics are tricks for memorizing things. For instance, the musical notes that lie between the lines on the treble clef spell the word FACE, which is easy to remember. Many people use the mnemonic "SOHCAHTOA" to remember the definitions of the sine, cosine, and tangent in trigonometry. I have my own modest offering, POFOSTITO, which I hope will make it into the mnemonics hall of fame. It's a way to avoid some of the most common problems with applying Newton's third law correctly:

Pair of
Opposite
Forces
Of the
Same
Type
Involving
Two
Objects

Example

Question: A book is lying on a table. What force is the Newton's-third-law partner of the earth's gravitational force on the book?
Answer: Newton's third law works like "B on A, A on B," so the partner must be the book's gravitational force pulling upward on the planet earth. Yes, there is such a force! No, it does not cause the earth to do anything noticeable.
Incorrect answer: The table's upward force on the book is the Newton's-third-law partner of the earth's gravitational force on the book.
✗ This answer violates two out of three of the commandments of POFOSTITO. The forces are not of the same type, because the table's upward force on the book is not gravitational. Also, three objects are involved instead of two: the book, the table, and the planet earth.

Example

Question: A person is pushing a box up a hill. What force is related by Newton's third law to the person's force on the box?
Answer: The box's force on the person.
Incorrect answer: The person's force on the box is opposed by friction, and also by gravity.
✗ This answer fails all three parts of the POFOSTITO test, the most obvious of which is that three forces are referred to instead of a pair.

Optional Topic: Newton's third law and action at a distance

Newton's third law is completely symmetric in the sense that neither force constitutes a delayed response to the other. Newton's third law does not even mention time, and the forces are supposed to agree at any given instant. This creates an interesting situation when it comes to noncontact forces. Suppose two people are holding magnets, and when one person waves or wiggles her magnet, the other person feels an effect on his. In this way they can send signals to each other from opposite sides of a wall, and if Newton's third law is correct, it would seem that the signals are transmitted instantly, with no time lag. The signals are indeed transmitted quite quickly, but experiments with electronically controlled magnets show that the signals do not leap the gap instantly: they travel at the same speed as light, which is an extremely high speed but not an infinite one.

Is this a contradiction to Newton's third law? Not really. According to current theories, there are no true noncontact forces. Action at a distance does not exist. Although it appears that the wiggling of one magnet affects the other with no need for anything to be in contact with anything, what really happens is that wiggling a magnet unleashes a shower of tiny particles called photons. The magnet shoves the photons out with a kick, and receives a kick in return, in strict obedience to Newton's third law. The photons fly out in all directions, and the ones that hit the other magnet then interact with it, again obeying Newton's third law.

Photons are nothing exotic, really. Light is made of photons, but our eyes receive such huge numbers of photons that we do not perceive them individually. The photons you would make by wiggling a magnet with your hand would be of a "color" that you cannot see, far off the red end of the rainbow. Book 6 in this series describes the evidence for the photon model of light.

Discussion questions

A. When you fire a gun, the exploding gases push outward in all directions, causing the bullet to accelerate down the barrel. What third-law pairs are involved? [Hint: Remember that the gases themselves are an object.]

B. Tam Anh grabs Sarah by the hand and tries to pull her. She tries to remain standing without moving. A student analyzes the situation as follows. "If Tam Anh's force on Sarah is greater than her force on him, he can get her to move. Otherwise, she'll be able to stay where she is." What's wrong with this analysis?

C. You hit a tennis ball against a wall. Explain any and all incorrect ideas in the following description of the physics involved: "According to Newton's third law, there has to be a force opposite to your force on the ball. The opposite force is the ball's mass, which resists acceleration, and also air resistance."

5.2 Classification and Behavior of Forces

One of the most basic and important tasks of physics is to classify the forces of nature. I have already referred informally to "types" of forces such as friction, magnetism, gravitational forces, and so on. Classification systems are creations of the human mind, so there is always some degree of arbitrariness in them. For one thing, the level of detail that is appropriate for a classification system depends on what you're trying to find out. Some linguists, the "lumpers," like to emphasize the similarities among languages, and a few extremists have even tried to find signs of similarities between words in languages as different as English and Chinese, lumping the world's languages into only a few large groups. Other linguists, the "splitters," might be more interested in studying the differences in pronunciation between English speakers in New York and Connecticut. The splitters call the lumpers sloppy, but the lumpers say that science isn't worthwhile unless it can find broad, simple patterns within the seemingly complex universe.

Scientific classification systems are also usually compromises between practicality and naturalness. An example is the question of how to classify flowering plants. Most people think that biological classification is about discovering new species, naming them, and classifying them in the class-order-family-genus-species system according to guidelines set long ago. In reality, the whole system is in a constant state of flux and controversy. One very practical way of classifying flowering plants is according to whether their petals are separate or joined into a tube or cone — the criterion is so clear that it can be applied to a plant seen from across the street. But here practicality conflicts with naturalness. For instance, the begonia has separate petals and the pumpkin has joined petals, but they are so similar in so many other ways that they are usually placed within the same order. Some taxonomists have come up with classification criteria that they claim correspond more naturally to the apparent relationships among plants, without having to make special exceptions, but these may be far less practical, requiring for instance the examination of pollen grains under an electron microscope.

In physics, there are two main systems of classification for forces. At this point in the course, you are going to learn one that is very practical and easy to use, and that splits the forces up into a relatively large number of types: seven very common ones that we'll discuss explicitly in this chapter, plus perhaps ten less important ones such as surface tension, which we will not bother with right now.

Professional physicists, however, are almost all obsessed with finding simple patterns, so recognizing as many as fifteen or twenty types of forces strikes them as distasteful and overly complex. Since about the year 1900, physics has been on an aggressive program to discover ways in which these many seemingly different types of forces arise from a smaller number of fundamental ones. For instance, when you press your hands together, the force that keeps them from passing through each other may seem to have nothing to do with electricity, but at the atomic level, it actually does arise from electrical repulsion between atoms. By about 1950, all the forces of nature had been explained as arising from four fundamental types of forces at the atomic and nuclear level, and the lumping-together process didn't stop there. By the 1960's the length of the list had been reduced to three, and some theorists even believe that they may be able to reduce it to two or one. Although the unification of the forces of nature is one of the most beautiful and important achievements of physics, it makes much more sense to start this course with the more practical and easy system of classification. The unified system of four forces will be one of the highlights of the end of your introductory physics sequence.

The practical classification scheme which concerns us now can be laid out in the form of the tree shown below. The most specific types of forces are shown at the tips of the branches, and it is these types of forces that are referred to in the POFOSTITO mnemonic. For example, electrical and magnetic forces belong to the same general group, but Newton's third law would never relate an electrical force to a magnetic force.

The broadest distinction is that between contact and noncontact forces, which has been discussed in the previous chapter. Among the contact forces, we distinguish between those that involve solids only and those that have to do with fluids, a term used in physics to include both gases and liquids. The terms "repulsive," "attractive," and "oblique" refer to the directions of the forces.

- Repulsive forces are those that tend to push the two participating objects away from each other. More specifically, a repulsive contact force acts perpendicular to the surfaces at which the two objects touch, and a repulsive noncontact force acts along the line between the two objects.

- Attractive forces pull the two objects toward one another, i.e. they act along the same line as repulsive forces, but in the opposite direction.

- Oblique forces are those that act at some other angle.

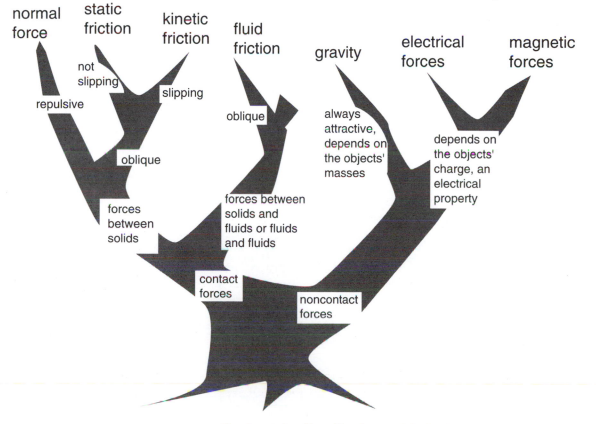

It should not be necessary to memorize this diagram by rote. It is better to reinforce your memory of this system by calling to mind your commonsense knowledge of certain ordinary phenomena. For instance, we know that the gravitational attraction between us and the planet earth will act even if our feet momentarily leave the ground, and that although magnets have mass and are affected by gravity, most objects that have mass are nonmagnetic.

This diagram is meant to be as simple as possible while including most of the forces we deal with in everyday life. If you were an insect, you would be much more interested in the force of surface tension, which allowed you to walk on water. I have not included the nuclear forces, which are responsible for holding the nuclei of atoms, because they are not evident in everyday life.

You should not be afraid to invent your own names for types of forces that do not fit into the diagram. For instance, the force that holds a piece of tape to the wall has been left off of the tree, and if you were analyzing a situation involving scotch tape, you would be absolutely right to refer to it by some commonsense name such as "sticky force."

On the other hand, if you are having trouble classifying a certain force, you should also consider whether it is a force at all. For instance, if someone asks you to classify the force that the earth has because of its rotation, you would have great difficulty creating a place for it on the diagram. That's because it's a type of motion, not a type of force!

Normal forces

A normal force, F_N, is a force that keeps one solid object from passing through another. "Normal" is simply a fancy word for "perpendicular," meaning that the force is perpendicular to the surface of contact. Intuitively, it seems the normal force magically adjusts itself to provide whatever force is needed to keep the objects from occupying the same space. If your muscles press your hands together gently, there is a gentle normal force. Press harder, and the normal force gets stronger. How does the normal force know how strong to be? The answer is that the harder you jam your hands together, the more compressed your flesh becomes. Your flesh is acting like a spring: more force is required to compress it more. The same is true when you push on a wall. The wall flexes imperceptibly in proportion to your force on it. If you exerted enough force, would it be possible for two objects to pass through each other? No, typically the result is simply to strain the objects so much that one of them breaks.

Gravitational forces

As we'll discuss in more detail later in the course, a gravitational force exists between any two things that have mass. In everyday life, the gravitational force between two cars or two people is negligible, so the only noticeable gravitational forces are the ones between the earth and various human-scale objects. We refer to these planet-earth-induced gravitational forces as weight forces, and as we have already seen, their magnitude is given by $|F_W|=mg$.

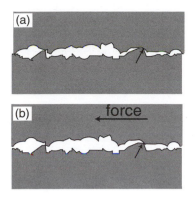

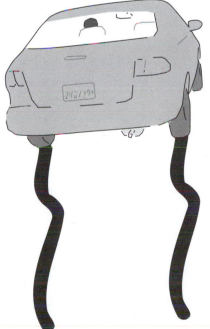

A model that correctly explains many properties of friction. The microscopic bumps and holes in two surfaces dig into each other, causing a frictional force.

Static friction: the tray doesn't slip on the waiter's fingers.

Kinetic friction: the car skids.

Static and kinetic friction

If you have pushed a refrigerator across a kitchen floor, you have felt a certain series of sensations. At first, you gradually increased your force on the refrigerator, but it didn't move. Finally, you supplied enough force to unstick the fridge, and there was a sudden jerk as the fridge started moving. Once the fridge is unstuck, you can reduce your force significantly and still keep it moving.

While you were gradually increasing your force, the floor's frictional force on the fridge increased in response. The two forces on the fridge canceled, and the fridge didn't accelerate. How did the floor know how to respond with just the right amount of force? The figures on the left show one possible *model* of friction that explains this behavior. (A scientific model is a description that we expect to be incomplete, approximate, or unrealistic in some ways, but that nevertheless succeeds in explaining a variety of phenomena.) Figure (a) shows a microscopic view of the tiny bumps and holes in the surfaces of the floor and the refrigerator. The weight of the fridge presses the two surfaces together, and some of the bumps in one surface will settle as deeply as possible into some of the holes in the other surface. In figure (b), your leftward force on the fridge has caused it to ride up a little higher on the bump in the floor labeled with a small arrow. Still more force is needed to get the fridge over the bump and allow it to start moving. Of course, this is occurring simultaneously at millions of places on the two surfaces.

Once you had gotten the fridge moving at constant speed, you found that you needed to exert less force on it. Since zero total force is needed to make an object move with constant velocity, the floor's rightward frictional force on the fridge has apparently decreased somewhat, making it easier for you to cancel it out. Our model also gives a plausible explanation for this fact: as the surfaces slide past each other, they don't have time to settle down and mesh with one another, so there is less friction.

Even though this model is intuitively appealing and fairly successful, it should not be taken too seriously, and in some situations it is misleading. For instance, fancy racing bikes these days are made with smooth tires that have no tread — contrary to what we'd expect from our model, this does not cause any decrease in friction. Machinists know that two very smooth and clean metal surfaces may stick to each other firmly and be very difficult to slide apart. This cannot be explained in our model, but makes more sense in terms of a model in which friction is described as arising from chemical bonds between the atoms of the two surfaces at their points of contact: very flat surfaces allow more atoms to come in contact.

Since friction changes its behavior dramatically once the surfaces come unstuck, we define two separate types of frictional forces. *Static friction* is friction that occurs between surfaces that are not slipping over each other. Slipping surfaces experience *kinetic friction*. "Kinetic" means having to do with motion. The forces of static and kinetic friction, notated F_s and F_k, are always parallel to the surface of contact between the two objects.

1. When a baseball player slides in to a base, is the friction static, or kinetic?
2. A mattress stays on the roof of a slowly accelerating car. Is the friction static or kinetic?
3. Does static friction create heat? Kinetic friction?

The maximum possible force of static friction depends on what kinds of surfaces they are, and also on how hard they are being pressed together. The approximate mathematical relationships can be expressed as follows:

$$F_s = -F_{applied}, \text{ when } |F_{applied}| < \mu_s |F_N| \ ,$$

where μ_s is a unitless number, called the coefficient of static friction, which depends on what kinds of surfaces they are. The maximum force that static friction can supply, $\mu_s |F_N|$, represents the boundary between static and kinetic friction. It depends on the normal force, which is numerically equal to whatever force is pressing the two surfaces together. In terms of our model, if the two surfaces are being pressed together more firmly, a greater sideways force will be required in order to make the irregularities in the surfaces ride up and over each other.

Note that just because we use an adjective such as "applied" to refer to a force, that doesn't mean that there is some special type of force called the "applied force." The applied force could be any type of force, or it could be the sum of more than one force trying to make an object move.

The force of kinetic friction on each of the two objects is in the direction that resists the slippage of the surfaces. Its magnitude is usually well approximated as

$$|F_k| = \mu_k |F_N|$$

where μ_k is the coefficient of kinetic friction. Kinetic friction is usually more or less independent of velocity.

We choose a coordinate system in which the applied force, i.e. the force trying to move the objects, is positive. The friction force is then negative, since it is in the opposite direction. As you increase the applied force, the force of static friction increases to match it and cancel it out, until the maximum force of static friction is surpassed. The surfaces then begin slipping past each other, and the friction force becomes smaller in absolute value.

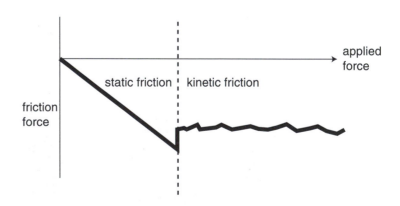

(1) It's kinetic friction, because her uniform is sliding over the dirt. (2) It's static friction, because even though the two surfaces are moving relative to the landscape, they're not slipping over each other. (3) Only kinetic friction creates heat, as when you rub your hands together. If you move your hands up and down together without sliding them across each other, no heat is produced by the static friction.

134   Chapter 5   Analysis of Forces

Self-Check

Can a frictionless surface exert a normal force? Can a frictional force exist without a normal force?

If you try to accelerate or decelerate your car too quickly, the forces between your wheels and the road become too great, and they begin slipping. This is not good, because kinetic friction is weaker than static friction, resulting in less control. Also, if this occurs while you are turning, the car's handling changes abruptly because the kinetic friction force is in a different direction than the static friction force had been: contrary to the car's direction of motion, rather than contrary to the forces applied to the tire.

Most people respond with disbelief when told of the experimental evidence that both static and kinetic friction are approximately independent of the amount of surface area in contact. Even after doing a hands-on exercise with spring scales to show that it is true, many students are unwilling to believe their own observations, and insist that bigger tires "give more traction." In fact, the main reason why you would not want to put small tires on a big heavy car is that the tires would burst!

Although many people expect that friction would be proportional to surface area, such a proportionality would make predictions contrary to many everyday observations. A dog's feet, for example, have very little surface area in contact with the ground compared to a human's feet, and yet we know that a dog can often win a tug-of-war with a person.

The reason why a smaller surface area does not lead to less friction is that the force between the two surfaces is more concentrated, causing their bumps and holes to dig into each other more deeply.

Frictionless ice can certainly make a normal force, since otherwise a hockey puck would sink into the ice. Friction is not possible without a normal force, however: we can see this from the equation, or from common sense, e.g. while sliding down a rope you do not get any friction unless you grip the rope.

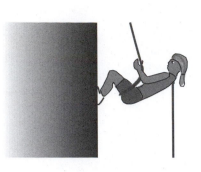

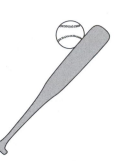

1. the cliff's normal force on the climber's feet

2. the track's static frictional force on the wheel

3. the ball's normal force on the bat

Self-Check

Find the direction of each of the forces in the figure above.

Fluid friction

Try to drive a nail into a waterfall and you will be confronted with the main difference between solid friction and fluid friction. Fluid friction is purely kinetic; there is no static fluid friction. The nail in the waterfall may tend to get dragged along by the water flowing past it, but it does not stick in the water. The same is true for gases such as air: recall that we are using the word "fluid" to include both gases and liquids.

Unlike solid kinetic friction, the force of fluid friction increases rapidly with velocity. In many cases, the force is approximately proportional to the square of the velocity,

$$F_{\text{fluid friction}} \propto c\rho A v^2 ,$$

where A is the cross-sectional area of the object, ρ is the density of the fluid, and c is a constant of proportionality that depends partly on the type of fluid and partly on how streamlined the object is.

Discussion questions

A. A student states that when he tries to push his refrigerator, the reason it won't move is because Newton's third law says there's an equal and opposite frictional force pushing back. After all, the static friction force is equal and opposite to the applied force. How would you convince him he is wrong?

B. Kinetic friction is usually more or less independent of velocity. However, inexperienced drivers tend to produce a jerk at the last moment of deceleration when they stop at a stop light. What does this tell you about the kinetic friction between the brake shoes and the brake drums?

(1) Normal forces are always perpendicular to the surface of contact, which means right or left in this figure. Normal forces are repulsive, so the cliff's force on the feet is to the right, i.e., away from the cliff. (2) Frictional forces are always parallel to the surface of contact, which means right or left in this figure. Static frictional forces are in the direction that would tend to keep the surfaces from slipping over each other. If the wheel was going to slip, its surface would be moving to the left, so the static frictional force on the wheel must be in the direction that would prevent this, i.e., to the right. This makes sense, because it is the static frictional force that accelerates the dragster. (3) Normal forces are always perpendicular to the surface of contact. In this diagram, that means either up and to the left or down and to the right. Normal forces are reulsive, so the ball is pushing the bat away from itself. Therefore the ball's force is down and to the right on this diagram.

C. Some of the following are correct descriptions of types of forces that could be added on as new branches of the classification tree. Others are not really types of forces, and still others are not force phenomena at all. In each case, decide what's going on, and if appropriate, figure out how you would incorporate them into the tree.

sticky force .... makes tape stick to things

opposite force ... the force that Newton's third law says relates to every force you make

flowing force ...... the force that water carries with it as it flows out of a hose

surface tension .. lets insects walk on water

horizontal force . a force that is horizontal

motor force .... the force that a motor makes on the thing it is turning

canceled force ... a force that is being canceled out by some other force

5.3   Analysis of Forces

Newton's first and second laws deal with the total of all the forces exerted on a specific object, so it is very important to be able to figure out what forces there are. Once you have focused your attention on one object and listed the forces on it, it is also helpful to describe all the corresponding forces that must exist according to Newton's third law. We refer to this as "analyzing the forces" in which the object participates.

Example

A barge is being pulled along a canal by teams of horses on the shores. Analyze all the forces in which the barge participates.

force acting on barge	force related to it by Newton's third law
ropes' forward normal forces on barge	barge's backward normal force on ropes
water's backward fluid friction force on barge	barge's forward fluid friction force on water
planet earth's downward gravitational force on barge	barge's upward gravitational force on earth
water's upward "floating" force on barge	barge's downward "floating" force on water

Here I've used the word "floating" force as an example of a sensible invented term for a type of force not classified on the tree in the previous section. A more formal technical term would be "hydrostatic force." Note how the pairs of forces are all structured as "A's force on B, B's force on A": ropes on barge and barge on ropes; water on barge and barge on water. Because all the forces in the left column are forces acting on the barge, all the forces in the right column are forces being exerted by the barge, which is why each entry in the column begins with "barge."

Often you may be unsure whether you have forgotten one of the forces. Here are three strategies for checking your list:

(1) See what physical result would come from the forces you've found so far. Suppose, for instance, that you'd forgotten the "floating" force on the barge in the example above. Looking at the forces you'd found, you would have found that there was a downward gravitational force on the barge which was not canceled by any upward force. The barge isn't supposed to sink, so you know you need to find a fourth, upward force.

(2) Another technique for finding missing forces is simply to go through the list of all the common types of forces and see if any of them apply.

(3) Make a drawing of the object, and draw a dashed boundary line around it that separates it from its environment. Look for points on the boundary where other objects come in contact with your object. This strategy guarantees that you'll find every contact force that acts on the object, although it won't help you to find non-contact forces.

The following is another example in which we can profit by checking against our physical intuition for what should be happening.

Example

As shown in the figure below, Cindy is rappelling down a cliff. Her downward motion is at constant speed, and she takes little hops off of the cliff, as shown by the dashed line. Analyze the forces in which she participates at a moment when her feet are on the cliff and she is pushing off.

force acting on Cindy	force related to it by Newton's third law
planet earth's downward gravitational force on Cindy	Cindy's upward gravitational force on earth
ropes upward frictional force on Cindy (her hand)	Cindy's downward frictional force on the rope
cliff's rightward normal force on Cindy	Cindy's leftward normal force on the cliff

The two vertical forces cancel, which is what they should be doing if she is to go down at a constant rate. The only horizontal force on her is the cliff's force, which is not canceled by any other force, and which therefore will produce an acceleration of Cindy to the right. This makes sense, since she is hopping off. (This solution is a little oversimplified, because the rope is slanting, so it also applies a small leftward force to Cindy. As she flies out to the right, the slant of the rope will increase, pulling her back in more strongly.)

I believe that constructing the type of table described in this section is the best method for beginning students. Most textbooks, however, prescribe a pictorial way of showing all the forces acting on an object. Such a picture is called a free-body diagram. It should not be a big problem if a future physics professor expects you to be able to draw such diagrams, because the conceptual reasoning is the same. You simply draw a picture of the object, with arrows representing the forces that are acting on it. Arrows representing contact forces are drawn from the point of contact, noncontact forces from the center of mass. Free-body diagrams do not show the equal and opposite forces exerted by the object itself.

Discussion questions

A. In the example of the barge going down the canal, I referred to a "floating" or "hydrostatic" force that keeps the boat from sinking. If you were adding a new branch on the force-classification tree to represent this force, where would it go?

B. A pool ball is rebounding from the side of the pool table. Analyze the forces in which the ball participates during the short time when it is in contact with the side of the table.

C. The earth's gravitational force on you, i.e. your weight, is always equal to mg, where m is your mass. So why can you get a shovel to go deeper into the ground by jumping onto it? Just because you're jumping, that doesn't mean your mass or weight is any greater, does it?

Discussion question C.

5.4   Transmission of Forces by Low-Mass Objects

You're walking your dog. The dog wants to go faster than you do, and the leash is taut. Does Newton's third law guarantee that your force on your end of the leash is equal and opposite to the dog's force on its end? If they're not exactly equal, is there any reason why they should be approximately equal?

If there was no leash between you, and you were in direct contact with the dog, then Newton's third law would apply, but Newton's third law cannot relate your force on the leash to the dog's force on the leash, because that would involve three separate objects. Newton's third law only says that your force on the leash is equal and opposite to the leash's force on you,

$$F_{yL} = -F_{Ly} \ ,$$

and that the dog's force on the leash is equal and opposite to its force on the dog

$$F_{dL} = -F_{Ld} \ .$$

Still, we have a strong intuitive expectation that whatever force we make on our end of the leash is transmitted to the dog, and vice-versa. We can analyze the situation by concentrating on the forces that act on the leash, F_{dL} and F_{yL}. According to Newton's second law, these relate to the leash's mass and acceleration:

$$F_{dL} + F_{yL} = m_L a_L \ .$$

The leash is far less massive then any of the other objects involved, and if m_L is very small, then apparently the total force on the leash is also very small, $F_{dL} + F_{yL} \approx 0$, and therefore

$$F_{dL} \approx -F_{yL} \ .$$

Thus even though Newton's third law does not apply directly to these two forces, we can approximate the low-mass leash as if it was not intervening between you and the dog. It's at least approximately as if you and the dog were acting directly on each other, in which case Newton's third law would have applied.

In general, low-mass objects can be treated approximately as if they simply transmitted forces from one object to another. This can be true for strings, ropes, and cords, and also for rigid objects such as rods and sticks.

If you look at a piece of string under a magnifying glass as you pull on the ends more and more strongly, you will see the fibers straightening and becoming taut. Different parts of the string are apparently exerting forces

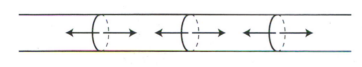

If we imagine dividing a taut rope up into small segments, then any segment has forces pulling outward on it at each end. If the rope is of negligible mass, then all the forces equal +T or -T, where T, the tension, is a single number.

on each other. For instance, if we think of the two halves of the string as two objects, then each half is exerting a force on the other half. If we imagine the string as consisting of many small parts, then each segment is transmitting a force to the next segment, and if the string has very little mass, then all the forces are equal in magnitude. We refer to the magnitude of the forces as the tension in the string, T. Although the tension is measured in units of Newtons, it is not itself a force. There are many forces within the string, some in one direction and some in the other direction, and their magnitudes are only approximately equal. The concept of tension only makes sense as a general, approximate statement of how big all the forces are.

If a rope goes over a pulley or around some other object, then the tension throughout the rope is approximately equal so long as there is not too much friction. A rod or stick can be treated in much the same way as a string, but it is possible to have either compression or tension.

Since tension is not a type of force, the force exerted by a rope on some other object must be of some definite type such as static friction, kinetic friction, or a normal force. If you hold your dog's leash with your hand through the loop, then the force exerted by the leash on your hand is a normal force: it is the force that keeps the leash from occupying the same space as your hand. If you grasp a plain end of a rope, then the force between the rope and your hand is a frictional force.

A more complex example of transmission of forces is the way a car accelerates. Many people would describe the car's engine as making the force that accelerates the car, but the engine is part of the car, so that's impossible: objects can't make forces on themselves. What really happens is that the engine's force is transmitted through the transmission to the axles, then through the tires to the road. By Newton's third law, there will thus be a forward force from the road on the tires, which accelerates the car.

Discussion question

When you step on the gas pedal, is your foot's force being transmitted in the sense of the word used in this section?

5.5   Objects Under Strain

A string lengthens slightly when you stretch it. Similarly, we have already discussed how an apparently rigid object such as a wall is actually flexing when it participates in a normal force. In other cases, the effect is more obvious. A spring or a rubber band visibly elongates when stretched.

Common to all these examples is a change in shape of some kind: lengthening, bending, compressing, etc. The change in shape can be measured by picking some part of the object and measuring its position, x. For concreteness, let's imagine a spring with one end attached to a wall. When no force is exerted, the unfixed end of the spring is at some position x_0. If a force acts at the unfixed end, its position will change to some new value of x. The more force, the greater the departure of x from x_0.

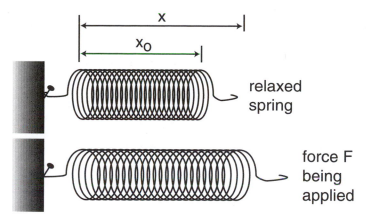

relaxed spring

force F being applied

Back in Newton's time, experiments like this were considered cutting-edge research, and his contemporary Hooke is remembered today for doing them and for coming up with a simple mathematical generalization called Hooke's law:

$$F \approx k(x-x_0) \quad \text{[force required to stretch a spring; valid for small forces only] .}$$

Here k is a constant, called the spring constant, that depends on how stiff the object is. If too much force is applied, the spring exhibits more complicated behavior, so the equation is only a good approximation if the force is sufficiently small. Usually when the force is so large that Hooke's law is a bad approximation, the force ends up permanently bending or breaking the spring.

Although Hooke's law may seem like a piece of trivia about springs, it is actually far more important than that, because all solid objects exert Hooke's-law behavior over some range of sufficiently small forces. For example, if you push down on the hood of a car, it dips by an amount that is directly proportional to the force. (But the car's behavior would not be as mathematically simple if you dropped a boulder on the hood!)

Discussion questions

A car is connected to its axles through big, stiff springs called shock absorbers, or "shocks." Although we've discussed Hooke's law above only in the case of stretching a spring, a car's shocks are continually going through both stretching and compression. In this situation, how would you interpret the positive and negative signs in Hooke's law?

5.6 Simple Machines: The Pulley

Even the most complex machines, such as cars or pianos, are built out of certain basic units called *simple machines*. The following are some of the main functions of simple machines:

transmitting a force: The chain on a bicycle transmits a force from the crank set to the rear wheel.

changing the direction of a force: If you push down on a seesaw, the other end goes up.

changing the speed and precision of motion: When you make the "come here" motion, your biceps only moves a couple of centimeters where it attaches to your forearm, but your arm moves much farther and more rapidly.

changing the amount of force: A lever or pulley can be used to increase or decrease the amount of force.

You are now prepared to understand one-dimensional simple machines, of which the pulley is the main example.

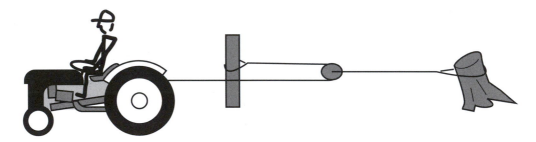

Example: a pulley

Question: Farmer Bill says this pulley arrangement doubles the force of his tractor. Is he just a dumb hayseed, or does he know what he's doing?

Solution: To use Newton's first law, we need to pick an object and consider the sum of the forces on it. Since our goal is to relate the tension in the part of the cable attached to the stump to the tension in the part attached to the tractor, we should pick an object to which both those cables are attached, i.e. the pulley itself. As discussed in section 5.4, the tension in a string or cable remains approximately constant as it passes around a pulley, provided that there is not too much friction. There are therefore two leftward forces acting on the pulley, each equal to the force exerted by the tractor. Since the acceleration of the pulley is essentially zero, the forces on it must be canceling out, so the rightward force of the pulley-stump cable on the pulley must be double the force exerted by the tractor. Yes, Farmer Bill knows what he's talking about.

Summary

Selected Vocabulary

repulsive	describes a force that tends to push the two participating objects apart
attractive	describes a force that tends to pull the two participating objects together
oblique	describes a force that acts at some other angle, one that is not a direct repulsion or attraction
normal force	the force that keeps two objects from occupying the same space
static friction	a friction force between surfaces that are not slipping past each other
kinetic friction	a friction force between surfaces that are slipping past each other
fluid	a gas or a liquid
fluid friction	a friction force in which at least one of the object is is a fluid
spring constant	the constant of proportionality between force and elongation of a spring or other object under strain

Notation

F_N	a normal force
F_s	a static frictional force
F_k	a kinetic frictional force
μ_s	the coefficient of static friction; the constant of proportionality between the maximum static frictional force and the normal force; depends on what types of surfaces are involved
μ_k	the coefficient of kinetic friction; the constant of proportionality between the kinetic frictional force and the normal force; depends on what types of surfaces are involved
k	the spring constant; the constant of proportionality between the force exerted on an object and the amount by which the object is lengthened or compressed

Summary

Newton's third law states that forces occur in equal and opposite pairs. If object A exerts a force on object B, then object B must simultaneously be exerting an equal and opposite force on object A. Each instance of Newton's third law involves exactly two objects, and exactly two forces, which are of the same type.

There are two systems for classifying forces. We are presently using the more practical but less fundamental one. In this system, forces are classified by whether they are repulsive, attractive, or oblique; whether they are contact or noncontact forces; and whether the two objects involved are solids or fluids.

Static friction adjusts itself to match the force that is trying to make the surfaces slide past each other, until the maximum value is reached,

$$|F_s| < \mu_s |F_N| \quad .$$

Once this force is exceeded, the surfaces slip past one another, and kinetic friction applies,

$$|F_k| = \mu_k |F_N| \quad .$$

Both types of frictional force are nearly independent of surface area, and kinetic friction is usually approximately independent of the speed at which the surfaces are slipping.

A good first step in applying Newton's laws of motion to any physical situation is to pick an object of interest, and then to list all the forces acting on that object. We classify each force by its type, and find its Newton's-third-law partner, which is exerted by the object on some other object.

When two objects are connected by a third low-mass object, their forces are transmitted to each other nearly unchanged.

Objects under strain always obey Hooke's law to a good approximation, as long as the force is small. Hooke's law states that the stretching or compression of the object is proportional to the force exerted on it,

$$F \approx k(x - x_o) \quad .$$

Homework Problems

1. A little old lady and a pro football player collide head-on. Compare their forces on each other, and compare their accelerations. Explain.

2. The earth is attracted to an object with a force equal and opposite to the force of the earth on the object. If this is true, why is it that when you drop an object, the earth does not have an acceleration equal and opposite to that of the object?

3. When you stand still, there are two forces acting on you, the force of gravity (your weight) and the normal force of the floor pushing up on your feet. Are these forces equal and opposite? Does Newton's third law relate them to each other? Explain.

In problems 4-8, analyze the forces using a table in the format shown in section 5.3. Analyze the forces in which the italicized object participates.

4. A *magnet* is stuck underneath a parked car.

5. Analyze two examples of *objects* at rest relative to the earth that are being kept from falling by forces other than the normal force. Do not use objects in outer space, and do not duplicate problem 4 or 8.

6. A *person* is rowing a boat, with her feet braced. She is doing the part of the stroke that propels the boat, with the ends of the oars in the water (not the part where the oars are out of the water).

Problem 6.

7. A *farmer* is in a stall with a cow when the cow decides to press him against the wall, pinning him with his feet off the ground. Analyze the forces in which the farmer participates.

8. A propeller *plane* is cruising east at constant speed and altitude.

9 . Today's tallest buildings are really not that much taller than the tallest buildings of the 1940s. One big problem with making an even taller skyscraper is that every elevator needs its own shaft running the whole height of the building. So many elevators are needed to serve the building's thousands of occupants that the elevator shafts start taking up too much of the space within the building. An alternative is to have elevators that can move both horizontally and vertically: with such a design, many elevator cars can share a few shafts, and they don't get in each other's way too much because they can detour around each other. In this design, it becomes impossible to hang the cars from cables, so they would instead have to ride on rails which they grab onto with wheels. Friction would keep them from slipping. The figure shows such a frictional elevator in its vertical travel mode. (The wheels on the bottom are for when it needs to switch to horizontal motion.) (a✔) If the coefficient of static friction between rubber and steel is μ_s, and the maximum mass of the car plus its passengers is M, how much force must there be pressing each wheel against the rail in order to keep the car from slipping? (Assume the car is not accelerating.) (b) Show that your result has physically reasonable behavior with respect to μ_s. In other words, if there was less friction, would the wheels need to be pressed more firmly or less firmly? Does your equation behave that way?

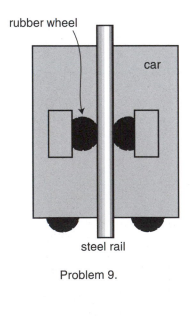

rubber wheel

car

steel rail

Problem 9.

S   A solution is given in the back of the book.
✓   A computerized answer check is available.
★   A difficult problem.
∫   A problem that requires calculus.

Problem 10.

10. Unequal masses M and m are suspended from a pulley as shown in the figure.

(a) Analyze the forces in which mass m participates, using a table the format shown in section 5.3. [The forces in which the other masses participate will of course be similar, but not numerically the same.]

(b✓) Find the magnitude of the accelerations of the two masses. [Hints: (1) Pick a coordinate system, and use positive and negative signs consistently to indicate the directions of the forces and accelerations. (2) The two accelerations of the two masses have to be equal in magnitude but of opposite signs, since one side eats up rope at the same rate at which the other side pays it out. (3) You need to apply Newton's second law twice, once to each mass, and then solve the two equations for the unknowns: the acceleration, a, and the tension in the rope, T.]

(c) Many people expect that in the special case of $M=m$, the two masses will naturally settle down to an equilibrium position side by side. Based on your answer from part (b), is this correct?

11. A tugboat of mass m pulls a ship of mass M, accelerating it. The speeds are low enough that you can ignore fluid friction acting on their hulls, although there will of course need to be fluid friction acting on the tug's propellers.

(a) Analyze the forces in which the tugboat participates, using a table in the format shown in section 5.3. Don't worry about vertical forces.

(b) Do the same for the ship.

(c✓) Assume now that water friction on the two vessels' hulls is negligible. If the force acting on the tug's propeller is F, what is the tension, T, in the cable connecting the two ships? [Hint: Write down two equations, one for Newton's second law applied to each object. Solve these for the two unknowns T and a.]

(d) Interpret your answer in the special cases of $M=0$ and $M=\infty$.

12. Explain why it wouldn't make sense to have kinetic friction be stronger than static friction.

13. In the system shown in the figure, the pulleys on the left and right are fixed, but the pulley in the center can move to the left or right. The two masses are identical. Show that the mass on the left will have an upward acceleration equal to $g/5$.

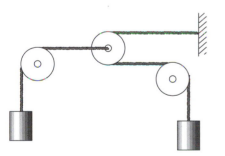

Problem 13.

14 S. The figure shows two different ways of combining a pair of identical springs, each with spring constant k. We refer to the top setup as parallel, and the bottom one as a series arrangement. (a) For the parallel arrangement, analyze the forces acting on the connector piece on the left, and then use this analysis to determine the equivalent spring constant of the whole setup. Explain whether the combined spring constant should be interpreted as being stiffer or less stiff. (b) For the series arrangement, analyze the forces acting on each spring and figure out the same things.

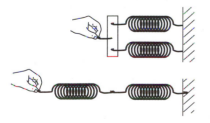

Problem 14.

15. Generalize the results of problem 14 to the case where the two spring constants are unequal.

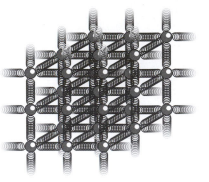

Problem 17.

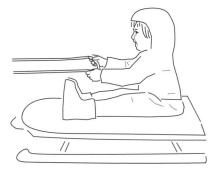

Problem 19.

16 S. (a) Using the solution of problem 14, which is given in the back of the book, predict how the spring constant of a fiber will depend on its length and cross-sectional area. (b) The constant of proportionality is called the Young's modulus, E, and typical values of the Young's modulus are about 10^{10} to 10^{11}. What units would the Young's modulus have in the SI (meter-kilogram-second) system?

17. This problem depends on the results of problems 14 and 16, whose solutions are in the back of the book. When atoms form chemical bonds, it makes sense to talk about the spring constant of the bond as a measure of how "stiff" it is. Of course, there aren't really little springs — this is just a mechanical model. The purpose of this problem is to estimate the spring constant, k, for a single bond in a typical piece of solid matter. Suppose we have a fiber, like a hair or a piece of fishing line, and imagine for simplicity that it is made of atoms of a single element stacked in a cubical manner, as shown in the figure, with a center-to-center spacing b. A typical value for b would be about 10^{-10} m. (a) Find an equation for k in terms of b, and in terms of the Young's modulus, E, defined in problem 16 and its solution. (b) Estimate k using the numerical data given in problem 16. (c) Suppose you could grab one of the atoms in a diatomic molecule like H_2 or O_2, and let the other atom hang vertically below it. Does the bond stretch by any appreciable fraction due to gravity?

18 S. In each case, identify the force that causes the acceleration, and give its Newton's-third-law partner. Describe the effect of the partner force. (a) A swimmer speeds up. (b) A golfer hits the ball off of the tee. (c) An archer fires an arrow. (d) A locomotive slows down.

19. Ginny has a plan. She is going to ride her sled while her dog Foo pulls her. However, Ginny hasn't taken physics, so there may be a problem: she may slide right off the sled when Foo starts pulling. (a) Analyze all the forces in which Giny participates, making a table as in section 5.3. (b) Analyze all the forces in which the sled participates. (c✓) The sled has mass m, and Ginny has mass M. The coefficient of static friction between the sled and the snow is μ_1, and μ_2 is the corresponding quantity for static friction between the sled and her snow pants. Ginny must have a certain minimum mass so that she will not slip off the sled. Find this in terms of the other three variables. (d) Under what conditions will there be no solution for M?

20 S. The second example in section 5.1 involves a person pushing a box up a hill. The incorrect answer describes three forces. For each of these three forces, give the force that it is related to by Newton's third law, and state the type of force.

21. The example in section 5.6 describes a force-doubling setup involving a pulley. Make up a more complicated arrangement, using more than one pullry, that would multiply the force by a factor greater than two.

22. Pick up a heavy object such as a backpack or a chair, and stand on a bathroom scale. Shake the object up and down. What do you observe? Interpret your observations in terms of Newton's third law.

23✓. A cop investigating the scene of an accident measures the length L of a car's skid marks in order to find out its speed v at the beginning of the skid. Express v in terms of L and any other relevant variables.

24. The following reasoning leads to an apparent paradox; explain what's wrong with the logic. A baseball player hits a ball. The ball and the bat spend a fraction of a second in contact. During that time they're moving together, so their accelerations must be equal. Newton's third law says that their forces on each other are also equal. But $a=F/m$, so how can this be, since their masses are unequal? (Note that the paradox isn't resolved by considering the force of the batter's hands on the bat. Not only is this force very small compared to the ball-bat force, but the batter could have just thrown the bat at the ball.)

Vectors are used in aerial navigation.

7 Vectors

7.1 Vector Notation

The idea of components freed us from the confines of one-dimensional physics, but the component notation can be unwieldy, since every one-dimensional equation has to be written as a set of three separate equations in the three-dimensional case. Newton was stuck with the component notation until the day he died, but eventually someone sufficiently lazy and clever figured out a way of abbreviating three equations as one.

(a)	$\vec{F}_{\text{A on B}} = -\vec{F}_{\text{B on A}}$	stands for	$F_{\text{A on B},x} = -F_{\text{B on A},x}$ $F_{\text{A on B},y} = -F_{\text{B on A},y}$ $F_{\text{A on B},z} = -F_{\text{B on A},z}$
(b)	$\vec{F}_{\text{total}} = \vec{F}_1 + \vec{F}_2 + ...$	stands for	$F_{\text{total},x} = F_{1,x} + F_{2,x} + ...$ $F_{\text{total},y} = F_{1,y} + F_{2,y} + ...$ $F_{\text{total},z} = F_{1,z} + F_{2,z} + ...$
(c)	$\vec{a} = \dfrac{\Delta \vec{v}}{\Delta t}$	stands for	$a_x = \Delta v_x / \Delta t$ $a_y = \Delta v_y / \Delta t$ $a_z = \Delta v_z / \Delta t$

Example (a) shows both ways of writing Newton's third law. Which would you rather write?

The idea is that each of the algebra symbols with an arrow written on top, called a *vector*, is actually an abbreviation for three different numbers, the x, y, and z components. The three components are referred to as the components of the vector, e.g. F_x is the x component of the vector $\vec{F}$. The notation with an arrow on top is good for handwritten equations, but is unattractive in a printed book, so books use boldface, **F**, to represent vectors. After this point, I'll use boldface for vectors throughout this book.

In general, the vector notation is useful for any quantity that has both an amount and a direction in space. Even when you are not going to write any actual vector notation, the concept itself is a useful one. We say that force and velocity, for example, are vectors. A quantity that has no direction in space, such as mass or time, is called a *scalar*. The amount of a vector quantity is called its *magnitude*. The notation for the magnitude of a vector **A** is $|A|$, like the absolute value sign used with scalars.

Often, as in example (b), we wish to use the vector notation to represent adding up all the x components to get a total x component, etc. The plus sign is used between two vectors to indicate this type of component-by-component addition. Of course, vectors are really triplets of numbers, not numbers, so this is not the same as the use of the plus sign with individual numbers. But since we don't want to have to invent new words and symbols for this operation on vectors, we use the same old plus sign, and the same old addition-related words like "add," "sum," and "total." Combining vectors this way is called *vector addition*.

Similarly, the minus sign in example (a) was used to indicate negating each of the vector's three components individually. The equals sign is used to mean that all three components of the vector on the left side of an equation are the same as the corresponding components on the right.

Example (c) shows how we abuse the division symbol in a similar manner. When we write the vector $\Delta \boldsymbol{v}$ divided by the scalar Δt, we mean the new vector formed by dividing each one of the velocity components by Δt.

It's not hard to imagine a variety of operations that would combine vectors with vectors or vectors with scalars, but only four of them are required in order to express Newton's laws:

operation	definition
vector + vector	Add component by component to make a new set of three numbers.
vector - vector	Subtract component by component to make a new set of three numbers.
vector · scalar	Multiply each component of the vector by the scalar.
vector / scalar	Divide each component of the vector by the scalar.

As an example of an operation that is not useful for physics, there just aren't any useful physics applications for dividing a vector by another vector component by component. In optional section 7.5, we discuss in more detail the fundamental reasons why some vector operations are useful and others useless.

We can do algebra with vectors, or with a mixture of vectors and scalars in the same equation. Basically all the normal rules of algebra apply, but if you're not sure if a certain step is valid, you should simply translate it into three component-based equations and see if it works.

Example
Question: If we are adding two force vectors, **F+G**, is it valid to assume as in ordinary algebra that **F+G** is the same as **G+F**?
Answer: To tell if this algebra rule also applies to vectors, we simply translate the vector notation into ordinary algebra notation. In terms of ordinary numbers, the components of the vector **F+G** would be F_x+G_x, F_y+G_y, and F_z+G_z, which are certainly the same three numbers as G_x+F_x, G_y+F_y, and G_z+F_z. Yes, **F+G** is the same as **G+F**.

It is useful to define a symbol **r** for the vector whose components are x, y, and z, and a symbol Δr made out of Δx, Δy, and Δz.

Although this may all seem a little formidable, keep in mind that it amounts to nothing more than a way of abbreviating equations! Also, to keep things from getting too confusing the remainder of this chapter focuses mainly on the Δr vector, which is relatively easy to visualize.

Self-Check

Translate the equations $v_x=\Delta x/\Delta t$, $v_y=\Delta y/\Delta t$, and $v_z=\Delta z/\Delta t$ for motion with constant velocity into a single equation in vector notation.

$v=\Delta r/\Delta t$

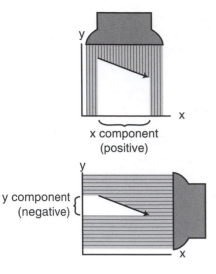

x component
(positive)

y component
(negative)

Drawing vectors as arrows

A vector in two dimensions can be easily visualized by drawing an arrow whose length represents its magnitude and whose direction represents its direction. The x component of a vector can then be visualized as the length of the shadow it would cast in a beam of light projected onto the x axis, and similarly for the y component. Shadows with arrowheads pointing back against the direction of the positive axis correspond to negative components.

In this type of diagram, the negative of a vector is the vector with the same magnitude but in the opposite direction. Multiplying a vector by a scalar is represented by lengthening the arrow by that factor, and similarly for division.

Self-Check

Given vector **Q** represented by an arrow below, draw arrows representing the vectors 1.5**Q** and —**Q**.

Q

Discussion Questions

A. Would it make sense to define a zero vector? Discuss what the zero vector's components, magnitude, and direction would be; are there any issues here? If you wanted to disqualify such a thing from being a vector, consider whether the system of vectors would be complete. For comparison, why is the ordinary number system (scalars) incomplete if you leave out zero? Does the same reasoning apply to vectors, or not?

B. You drive to your friend's house. How does the magnitude of your Δr vector compare with the distance you've added to the car's odometer?

7.2 Calculations with Magnitude and Direction

If you ask someone where Las Vegas is compared to Los Angeles, they are unlikely to say that the Δx is 290 km and the Δy is 230 km, in a coordinate system where the positive x axis is east and the y axis points north. They will probably say instead that it's 370 km to the northeast. If they were being precise, they might specify the direction as 38° counterclockwise from east. In two dimensions, we can always specify a vector's direction like this, using a single angle. A magnitude plus an angle suffice to specify everything about the vector. The following two examples show how we use trigonometry and the Pythagorean theorem to go back and forth between the x-y and magnitude-angle descriptions of vectors.

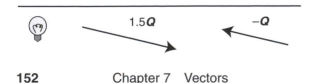

1.5**Q**   —**Q**

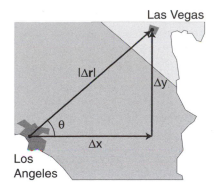

Las Vegas

Los Angeles

$|\Delta r|$

Δy

θ

Δx

Example: finding the magnitude and angle from the components
Question: Given that the $\Delta \mathbf{r}$ vector from LA to Las Vegas has Δx=290 km and Δy=230 km, how would we find the magnitude and direction of $\Delta \mathbf{r}$?

Solution: We find the magnitude of $\Delta \mathbf{r}$ from the Pythagorean theorem:

$$|\Delta r| = \sqrt{\Delta x^2 + \Delta y^2}$$
$$= 370 \text{ km}$$

We know all three sides of the triangle, so the angle θ can be found using any of the inverse trig functions. For example, we know the opposite and adjacent sides, so

$$\theta = \tan^{-1}\frac{\Delta y}{\Delta x}$$
$$= 38° \quad .$$

Example: finding the components from the magnitude and angle
Question: Given that the straight-line distance from Los Angeles to Las Vegas is 370 km, and that the angle θ in the figure is 38°, how can the x and y components of the $\Delta \mathbf{r}$ vector be found?

Solution: The sine and cosine of θ relate the given information to the information we wish to find:

$$\cos\theta = \frac{\Delta x}{|\Delta r|}$$

$$\sin\theta = \frac{\Delta y}{|\Delta r|}$$

Solving for the unknowns gives

$$\Delta x = |\Delta r|\cos\theta$$
$$= 290 \text{ km}$$
$$\Delta y = |\Delta r|\sin\theta$$
$$= 230 \text{ km}$$

The following example shows the correct handling of the plus and minus signs, which is usually the main cause of mistakes.

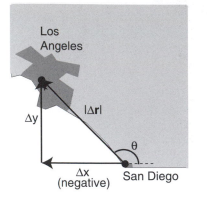

Example: negative components
Question: San Diego is 120 km east and 150 km south of Los Angeles. An airplane pilot is setting course from San Diego to Los Angeles. At what angle should she set her course, measured counterclockwise from east, as shown in the figure?
Solution: If we make the traditional choice of coordinate axes, with x pointing to the right and y pointing up on the map, then her Δx is negative, because her final x value is less than her initial x value. Her Δy is positive, so we have

$$\Delta x \quad = \text{-120 km}$$
$$\Delta y \quad = \text{150 km} \quad.$$

If we work by analogy with the previous example, we get

$$\theta \quad = \tan^{-1}\frac{\Delta y}{\Delta x}$$
$$= \tan^{-1}\left(-1.25\right)$$
$$= \text{-51°} \quad.$$

According to the usual way of defining angles in trigonometry, a negative result means an angle that lies clockwise from the x axis, which would have her heading for the Baja California. What went wrong? The answer is that when you ask your calculator to take the arctangent of a number, there are always two valid possibilities differing by 180°. That is, there are two possible angles whose tangents equal -1.25:

$$\tan 129° = \text{-1.25}$$
$$\tan \text{-51°} = \text{-1.25}$$

You calculator doesn't know which is the correct one, so it just picks one. In this case, the one it picked was the wrong one, and it was up to you to add 180° to it to find the right answer.

Discussion Question

In the example above, we dealt with *components* that were negative. Does it make sense to talk about positive and negative *vectors*?

7.3   Techniques for Adding Vectors

Addition of vectors given their components

The easiest type of vector addition is when you are in possession of the components, and want to find the components of their sum.

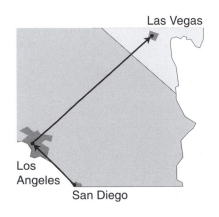

Example
Question: Given the Δx and Δy values from the previous examples, find the Δx and Δy from San Diego to Las Vegas.
Solution:

$$\Delta x_{\text{total}} \quad = \Delta x_1 + \Delta x_2$$
$$= -120 \text{ km} + 290 \text{ km}$$
$$= 170 \text{ km}$$
$$\Delta y_{\text{total}} \quad = \Delta y_1 + \Delta y_2$$
$$= 150 \text{ km} + 230 \text{ km}$$
$$= 380$$

Note how the signs of the *x* components take care of the westward and eastward motions, which partially cancel.

Addition of vectors given their magnitudes and directions

In this case, you must first translate the magnitudes and directions into components, and the add the components.

Graphical addition of vectors

Often the easiest way to add vectors is by making a scale drawing on a piece of paper. This is known as graphical addition, as opposed to the analytic techniques discussed previously.

Example
Question: Given the magnitudes and angles of the Δr vectors from San Diego to Los Angeles and from Los Angeles to Las Vegas, find the magnitude and angle of the Δr vector from San Diego to Las Vegas.
Solution: Using a protractor and a ruler, we make a careful scale drawing, as shown in the figure. A scale of 1 cm→100 km was chosen for this solution. With a ruler, we measure the distance from San Diego to Las Vegas to be 3.8 cm, which corresponds to 380 km. With a protractor, we measure the angle θ to be 71°.

Even when we don't intend to do an actual graphical calculation with a ruler and protractor, it can be convenient to diagram the addition of vectors in this way. With Δr vectors, it intuitively makes sense to lay the vectors tip-to-tail and draw the sum vector from the tail of the first vector to the tip of the second vector. We can do the same when adding other vectors such as force vectors.

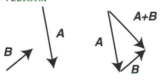

Vectors can be added graphically by placing them tip to tail, and then drawing a vector from the tail of the first vector to the tip of the second vector.

Self-Check

How would you subtract vectors graphically?

Discussion Questions

A. If you're doing *graphical* addition of vectors, does it matter which vector you start with and which vector you start from the other vector's tip?
B. If you add a vector with magnitude 1 to a vector of magnitude 2, what magnitudes are possible for the vector sum?
C. Which of these examples of vector addition are correct, and which are incorrect?

The difference $A{-}B$ is equivalent to $A{+}({-}B)$, which can be calculated graphically by reversing B to form $-B$, and then adding it to A.

7.4* Unit Vector Notation

When we want to specify a vector by its components, it can be cumbersome to have to write the algebra symbol for each component:

$$\Delta x = 290 \text{ km}, \Delta y = 230 \text{ km}$$

A more compact notation is to write

$$\Delta \boldsymbol{r} = (290 \text{ km})\hat{\boldsymbol{x}} + (230 \text{ km})\hat{\boldsymbol{y}} \quad ,$$

where the vectors $\hat{\boldsymbol{x}}$, $\hat{\boldsymbol{y}}$, and $\hat{\boldsymbol{z}}$, called the unit vectors, are defined as the vectors that have magnitude equal to 1 and directions lying along the x, y, and z axes. In speech, they are referred to as "x-hat" and so on.

A slightly different, and harder to remember, version of this notation is unfortunately more prevalent. In this version, the unit vectors are called $\hat{\boldsymbol{i}}$, $\hat{\boldsymbol{j}}$, and $\hat{\boldsymbol{k}}$:

$$\Delta \boldsymbol{r} = (290 \text{ km})\hat{\boldsymbol{i}} + (230 \text{ km})\hat{\boldsymbol{j}} \quad .$$

7.5* Rotational Invariance

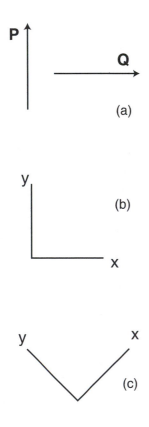

(a)

(b)

(c)

Let's take a closer look at why certain vector operations are useful and others are not. Consider the operation of multiplying two vectors component by component to produce a third vector:

$$
\begin{aligned}
R_x &= P_x Q_x \\
R_y &= P_y Q_y \\
R_z &= P_z Q_z
\end{aligned}
$$

As a simple example, we choose vectors **P** and **Q** to have length 1, and make them perpendicular to each other, as shown in figure (a). If we compute the result of our new vector operation using the coordinate system shown in (b), we find:

$$
\begin{aligned}
R_x &= 0 \\
R_y &= 0 \\
R_z &= 0
\end{aligned}
$$

The x component is zero because $P_x = 0$, the y component is zero because $Q_y = 0$, and the z component is of course zero because both vectors are in the x-y plane. However, if we carry out the same operations in coordinate system (c), rotated 45 degrees with respect to the previous one, we find

$$
\begin{aligned}
R_x &= 1/2 \\
R_y &= -1/2 \\
R_z &= 0
\end{aligned}
$$

The operation's result depends on what coordinate system we use, and since the two versions of R have different lengths (one being zero and the other nonzero), they don't just represent the same answer expressed in two different coordinate systems. Such an operation will never be useful in physics, because experiments show physics works the same regardless of which way we orient the laboratory building! The *useful* vector operations, such as addition and scalar multiplication, are rotationally invariant, i.e. come out the same regardless of the orientation of the coordinate system.

Summary

Selected Vocabulary

vector	a quantity that has both an amount (magnitude) and a direction in space
magnitude	the "amount" associated with a vector
scalar	a quantity that has no direction in space, only an amount

Notation

A	vector with components A_x, A_y, and A_z
$\vec{A}$	handwritten notation for a vector
$\lvert A \rvert$	the magnitude of vector A
r	the vector whose components are x, y, and z
Δr	the vector whose components are Δx, Δy, and Δz
$\hat{x}, \hat{y}, \hat{z}$	(optional topic) unit vectors; the vectors with magnitude 1 lying along the x, y, and z axes
$\hat{i}, \hat{j}, \hat{k}$	a harder to remember notation for the unit vectors

Standard Terminology Avoided in This Book

displacement vector	a name for the symbol Δr
speed	the magnitude of the velocity vector, i.e. the velocity stripped of any information about its direction

Summary

A vector is a quantity that has both a magnitude (amount) and a direction in space, as opposed to a scalar, which has no direction. The vector notation amounts simply to an abbreviation for writing the vector's three components.

In two dimensions, a vector can be represented either by its two components or by its magnitude and direction. The two ways of describing a vector can be related by trigonometry.

The two main operations on vectors are addition of a vector to a vector, and multiplication of a vector by a scalar.

Vector addition means adding the components of two vectors to form the components of a new vector. In graphical terms, this corresponds to drawing the vectors as two arrows laid tip-to-tail and drawing the sum vector from the tail of the first vector to the tip of the second one. Vector subtraction is performed by negating the vector to be subtracted and then adding.

Multiplying a vector by a scalar means multiplying each of its components by the scalar to create a new vector. Division by a scalar is defined similarly.

Homework Problems

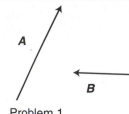

Problem 1.

1. The figure shows vectors A and B. Graphically calculate the following:

$$A+B, \quad A-B, \quad B-A, \quad -2B, \quad A-2B$$

No numbers are involved.

2. Phnom Penh is 470 km east and 250 km south of Bangkok. Hanoi is 60 km east and 1030 km north of Phnom Penh. (a) Choose a coordinate system, and translate these data into Δx and Δy values with the proper plus and minus signs. (b✓) Find the components of the Δr vector pointing from Bangkok to Hanoi.

3 ✓. If you walk 35 km at an angle 25° counterclockwise from east, and then 22 km at 230° counterclockwise from east, find the distance and direction from your starting point to your destination.

S   A solution is given in the back of the book.   ★ A difficult problem.
✓ A computerized answer check is available.   ∫ A problem that requires calculus.

THE HORSE IN MOTION.

Illustrated by
MUYBRIDGE.

8 Vectors and Motion

In 1872, capitalist and former California governor Leland Stanford asked photographer Eadweard Muybridge if he would work for him on a project to settle a $25,000 bet (a princely sum at that time). Stanford's friends were convinced that a galloping horse always had at least one foot on the ground, but Stanford claimed that there was a moment during each cycle of the motion when all four feet were in the air. The human eye was simply not fast enough to settle the question. In 1878, Muybridge finally succeeded in producing what amounted to a motion picture of the horse, showing conclusively that all four feet did leave the ground at one point. (Muybridge was a colorful figure in San Francisco history, and his acquittal for the murder of his wife's lover was considered the trial of the century in California.)

The losers of the bet had probably been influenced by Aristotelian reasoning, for instance the expectation that a leaping horse would lose horizontal velocity while in the air with no force to push it forward, so that it would be more efficient for the horse to run without leaping. But even for students who have converted wholeheartedly to Newtonianism, the relationship between force and acceleration leads to some conceptual difficulties, the main one being a problem with the true but seemingly absurd statement that an object can have an acceleration vector whose direction is not the same as the direction of motion. The horse, for instance, has nearly constant horizontal velocity, so its a_x is zero. But as anyone can tell you who has ridden a galloping horse, the horse accelerates up and down. The horse's acceleration vector therefore changes back and forth between the up and down directions, but is never in the same direction as the horse's motion. In this chapter, we will examine more carefully the properties of the velocity, acceleration, and force vectors. No new principles are introduced, but an attempt is made to tie things together and show examples of the power of the vector formulation of Newton's laws.

8.1 The Velocity Vector

For motion with constant velocity, the velocity vector is

$$\mathbf{v} = \Delta\mathbf{r}/\Delta t \qquad\text{[only for constant velocity] .}$$

The $\Delta\mathbf{r}$ vector points in the direction of the motion, and dividing it by the scalar Δt only changes its length, not its direction, so the velocity vector points in the same direction as the motion. When the velocity is not constant, i.e. when the x-t, y-t, and z-t graphs are not all linear, we use the slope-of-the-tangent-line approach to define the components v_x, v_y, and v_z, from which we assemble the velocity vector. Even when the velocity vector is not constant, it still points along the direction of motion.

Vector addition is the correct way to generalize the one-dimensional concept of adding velocities in relative motion, as shown in the following example:

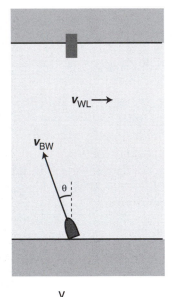

Example: velocity vectors in relative motion
Question: You wish to cross a river and arrive at a dock that is directly across from you, but the river's current will tend to carry you downstream. To compensate, you must steer the boat at an angle. Find the angle θ, given the magnitude, $|\mathbf{v}_{WL}|$, of the water's velocity relative to the land, and the maximum speed, $|\mathbf{v}_{BW}|$, of which the boat is capable relative to the water.
Solution: The boat's velocity relative to the land equals the vector sum of its velocity with respect to the water and the water's velocity with respect to the land,

$$\mathbf{v}_{BL} = \mathbf{v}_{BW} + \mathbf{v}_{WL} .$$

If the boat is to travel straight across the river, i.e. along the y axis, then we need to have $v_{BL,x}=0$. This x component equals the sum of the x components of the other two vectors,

$$v_{BL,x} = v_{BW,x} + v_{WL,x} ,$$

or

$$0 = -|\mathbf{v}_{BW}| \sin\theta + |\mathbf{v}_{WL}| .$$

Solving for θ, we find

$$\sin\theta = |\mathbf{v}_{WL}|/|\mathbf{v}_{BW}| ,$$

$$\theta = \sin^{-1}\frac{\left|\mathbf{v}_{WL}\right|}{\left|\mathbf{v}_{BW}\right|} .$$

8.2   The Acceleration Vector

When all three acceleration components are constant, i.e. when the v_x-t, v_y-t, and v_z-t graphs are all linear, we can define the acceleration vector as

$$a = \Delta v / \Delta t \qquad \text{[only for constant acceleration]},$$

which can be written in terms of initial and final velocities as

$$a = (v_f - v_i)/\Delta t \qquad \text{[only for constant acceleration]}.$$

If the acceleration is not constant, we define it as the vector made out of the a_x, a_y, and a_z components found by applying the slope-of-the-tangent-line technique to the v_x-t, v_y-t, and v_z-t graphs.

Now there are two ways in which we could have a nonzero acceleration. Either the magnitude or the direction of the velocity vector could change. This can be visualized with arrow diagrams as shown in the figure. Both the magnitude and direction can change simultaneously, as when a car accelerates while turning. Only when the magnitude of the velocity changes while its direction stays constant do we have a Δv vector and an acceleration vector along the same line as the motion.

(a) A change in the magnitude of the velocity vector implies an acceleration.

Self-Check

(1) In figure (a), is the object speeding up or slowing down? (2) What would the diagram look like if v_i was the same as v_f? (3) Describe how the Δv vector is different depending on whether an object is speeding up or slowing down.

If this all seems a little strange and abstract to you, you're not alone. It doesn't mean much to most physics students the first time someone tells them that acceleration is a vector, and that the acceleration vector does not have to be in the same direction as the velocity vector. One way to understand those statements better is to imagine an object such as an air freshener or a pair of fuzzy dice hanging from the rear-view mirror of a car. Such a hanging object, called a bob, constitutes an accelerometer. If you watch the bob as you accelerate from a stop light, you'll see it swing backward. The horizontal direction in which the bob tilts is opposite to the direction of the acceleration. If you apply the brakes and the car's acceleration vector points backward, the bob tilts forward.

After accelerating and slowing down a few times, you think you've put your accelerometer through its paces, but then you make a right turn. Surprise! Acceleration is a vector, and needn't point in the same direction as the velocity vector. As you make a right turn, the bob swings outward, to your left. That means the car's acceleration vector is to your right, perpen-

(b) A change in the direction of the velocity vector also produces a non-zero Δv vector, and thus a nonzero acceleration vector, $\Delta v / \Delta t$.

dicular to your velocity vector. A useful definition of an acceleration vector should relate in a systematic way to the actual physical effects produced by the acceleration, so a physically reasonable definition of the acceleration vector must allow for cases where it is not in the same direction as the motion.

Self-Check

In projectile motion, what direction does the acceleration vector have?

The following are two examples of force, velocity, and acceleration vectors in complex motion.

This figure shows outlines traced from the first, third, fifth, seventh, and ninth frames in Muybridge's series of photographs of the galloping horse. The estimated location of the horse's center of mass is shown with a circle, which bobs above and below the horizontal dashed line.

If we don't care about calculating velocities and accelerations in any particular system of units, then we can pretend that the time between frames is one unit. The horse's velocity vector as it moves from one point to the next can then be found simply by drawing an arrow to connect one position of the center of mass to the next. This produces a series of velocity vectors which alternate between pointing above and below horizontal.

The $\Delta \boldsymbol{v}$ vector is the vector which we would have to add onto one velocity vector in order to get the next velocity vector in the series. The $\Delta \boldsymbol{v}$ vector alternates between pointing down (around the time when the horse is in the air, b) and up (around the time when the horse has two feet on the ground, d).

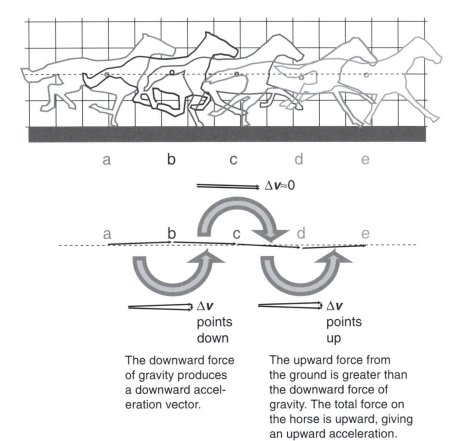

The downward force of gravity produces a downward acceleration vector.

The upward force from the ground is greater than the downward force of gravity. The total force on the horse is upward, giving an upward acceleration.

As we have already seen, the projectile has $a_x=0$ and $a_y=-g$, so the acceleration vector is pointing straight down.

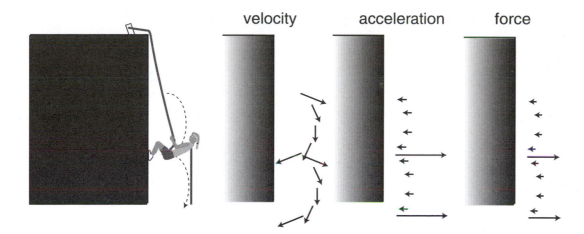

velocity   acceleration   force

In this example, the rappeller's velocity has long periods of gradual change interspersed with short periods of rapid change. These correspond to periods of small acceleration and force and periods of large acceleration and force.

Discussion Questions

A. When a car accelerates, why does a bob hanging from the rearview mirror swing toward the back of the car? Is it because a force throws it backward? If so, what force? Similarly, describe what happens in the other cases described above.

B. The following is a question commonly asked by students:

"Why does the force vector always have to point in the same direction as the acceleration vector? What if you suddenly decide to change your force on an object, so that your force is no longer pointing the same direction that the object is accelerating?"

What misunderstanding is demonstrated by this question? Suppose, for example, a spacecraft is blasting its rear main engines while moving forward, then suddenly begins firing its sideways maneuvering rocket as well. What does the student think Newton's laws are predicting?

8.3 The Force Vector and Simple Machines

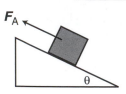

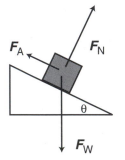

(a) The applied force $\mathbf{F}_A$ pushes the block up the frictionless ramp.

(b) Three forces act on the block. Their vector sum is zero.

(c) If the block is to move at constant velocity, Newton's first law says that the three force vectors acting on it must add up to zero. To perform vector addition, we put the vectors tip to tail, and in this case we are adding three vectors, so each one's tail goes against the tip of the previous one. Since they are supposed to add up to zero, the third vector's tip must come back to touch the tail of the first vector. They form a triangle, and since the applied force is perpendicular to the normal force, it is a right triangle.

Force is relatively easy to intuit as a vector. The force vector points in the direction in which it is trying to accelerate the object it is acting on.

Since force vectors are so much easier to visualize than acceleration vectors, it is often helpful to first find the direction of the (total) force vector acting on an object, and then use that information to determine the direction of the acceleration vector. Newton's second law, $\mathbf{F}_{total} = m\mathbf{a}$, tells us that the two must be in the same direction.

An important application of force vectors is to analyze the forces acting in two-dimensional mechanical systems, as in the following example.

Example: pushing a block up a ramp
Question: Figure (a) shows a block being pushed up a friction-less ramp at constant speed by an applied force $\mathbf{F}_A$. How much force is required, in terms of the block's mass, m, and the angle of the ramp, θ?
Solution: Figure (b) shows the other two forces acting on the block: a normal force, $\mathbf{F}_N$, created by the ramp, and the weight force, $\mathbf{F}_W$, created by the earth's gravity. Because the block is being pushed up at constant speed, it has zero acceleration, and the total force on it must be zero. From figure (c), we find

$$|\mathbf{F}_A| = |\mathbf{F}_W| \sin \theta$$
$$= mg \sin \theta \ .$$

Since the sine is always less than one, the applied force is always less than mg, i.e. pushing the block up the ramp is easier than lifting it straight up. This is presumably the principle on which the pyramids were constructed: the ancient Egyptians would have had a hard time applying the forces of enough slaves to equal the full weight of the huge blocks of stone.

Essentially the same analysis applies to several other simple machines, such as the wedge and the screw.

Discussion Questions

A. The figure shows a block being pressed diagonally upward against a wall, causing it to slide up the wall. Analyze the forces involved, including their directions.

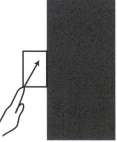

Discussion question A.

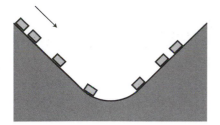

Discussion question C.

B. The figure shows a roller coaster car rolling down and then up under the influence of gravity. Sketch the car's velocity vectors and acceleration vectors. Pick an interesting point in the motion and sketch a set of force vectors acting on the car whose vector sum could have resulted in the right acceleration vector.

8.4 ∫ Calculus With Vectors

The definitions of the velocity and acceleration components given in chapter 6 can be translated into calculus notation as

$$v = \frac{dx}{dt}\hat{x} + \frac{dy}{dt}\hat{y} + \frac{dz}{dt}\hat{z}$$

and

$$a = \frac{dv_x}{dt}\hat{x} + \frac{dv_y}{dt}\hat{y} + \frac{dv_z}{dt}\hat{z} \quad .$$

To make the notation less cumbersome, we generalize the concept of the derivative to include derivatives of vectors, so that we can abbreviate the above equations as

$$v = \frac{d\mathbf{r}}{dt}$$

and

$$a = \frac{d\mathbf{v}}{dt} \quad .$$

In words, to take the derivative of a vector, you take the derivatives of its components and make a new vector out of those. This definition means that the derivative of a vector function has the familiar properties

$$\frac{d(cf)}{dt} = c\frac{df}{dt} \qquad\qquad [c \text{ is a constant}]$$

and

$$\frac{d(f+g)}{dt} = \frac{df}{dt} + \frac{dg}{dt} \quad . \qquad\qquad [c \text{ is a constant}]$$

The integral of a vector is likewise defined as integrating component by component.

Example

Question: Two objects have positions as functions of time given by the equations

$$r_1 = 3t^2 \hat{x} + t\hat{y}$$

and

$$r_2 = 3t^4 \hat{x} + t\hat{y} \quad .$$

Find both objects' accelerations using calculus. Could either answer have been found without calculus?

Solution: Taking the first derivative of each component, we find

$$v_1 = 6t\hat{x} + \hat{y}$$

$$v_2 = 12t^3\hat{x} + \hat{y} \quad ,$$

and taking the derivatives again gives acceleration,

$$a_1 = 6\hat{x}$$

$$a_2 = 36t^2\hat{x} \quad .$$

The first object's acceleration could have been found without calculus, simply by comparing the x and y coordinates with the constant-acceleration equation $\Delta x = v_0\Delta t + \frac{1}{2}a\Delta t^2$. The second equation, however, isn't just a second-order polynomial in t, so the acceleration isn't constant, and we really did need calculus to find the corresponding acceleration.

Example: a fire-extinguisher stunt on ice

Question: Prof. Puerile smuggles a fire extinguisher into a skating rink. Climbing out onto the ice without any skates on, he sits down and pushes off from the wall with his feet, acquiring an initial velocity $v_o\hat{y}$. At $t=0$, he then discharges the fire extinguisher at a 45-degree angle so that it applies a force to him that is backward and to the left, i.e. along the negative y axis and the positive x axis. The fire extinguisher's force is strong at first, but then dies down according to the equation $|F|=b-ct$, where b and c are constants. Find the professor's velocity as a function of time.

Solution: Measured counterclockwise from the x axis, the angle of the force vector becomes 315°. Breaking the force down into x and y components, we have

$$F_x \quad = \quad |F| \cos 315°$$

$$= \quad \tfrac{1}{\sqrt{2}} (b-ct)$$

$$F_y \quad = \quad |F| \sin 315°$$

$$= \quad \tfrac{1}{\sqrt{2}} (-b+ct) \quad .$$

In unit vector notation, this is

$$F \quad = \quad \tfrac{1}{\sqrt{2}} (b-ct)\hat{x} + \tfrac{1}{\sqrt{2}} (-b+ct)\hat{y} \quad .$$

Newton's second law gives

$$a \quad = \quad F/m$$

$$= \quad \frac{b-ct}{\sqrt{2}\,m} \hat{x} + \frac{-b+ct}{\sqrt{2}\,m} \hat{y} \quad .$$

To find the velocity vector as a function of time, we need to integrate the acceleration vector with respect to time,

$$\mathbf{v} = \int \mathbf{a}\,dt$$

$$= \int \left(\frac{b-ct}{\sqrt{2}\,m}\hat{\mathbf{x}} + \frac{-b+ct}{\sqrt{2}\,m}\hat{\mathbf{y}} \right) dt$$

$$= \frac{1}{\sqrt{2}\,m} \int \left[(b-ct)\hat{\mathbf{x}} + (-b+ct)\hat{\mathbf{y}} \right] dt$$

A vector function can be integrated component by component, so this can be broken down into two integrals,

$$\mathbf{v} = \frac{\hat{\mathbf{x}}}{\sqrt{2}\,m} \int (b-ct)\,dt + \frac{\hat{\mathbf{y}}}{\sqrt{2}\,m} \int (-b+ct)\,dt$$

$$= \left(\frac{bt - \frac{1}{2}ct^2}{\sqrt{2}\,m} + \text{const. \#1} \right) \hat{\mathbf{x}}$$

$$+ \left(\frac{-bt + \frac{1}{2}ct^2}{\sqrt{2}\,m} + \text{const. \#2} \right) \hat{\mathbf{y}}$$

Here the physical significance of the two constants of integration is that they give the initial velocity. Constant #1 is therefore zero, and constant #2 must equal v_o. The final result is

$$\mathbf{v} = \frac{bt - \frac{1}{2}ct^2}{\sqrt{2}\,m}\hat{\mathbf{x}} + \left(\frac{-bt + \frac{1}{2}ct^2}{\sqrt{2}\,m} + v_o \right) \hat{\mathbf{y}} \quad.$$

Summary

The velocity vector points in the direction of the object's motion. Relative motion can be described by vector addition of velocities.

The acceleration vector need not point in the same direction as the object's motion. We use the word "acceleration" to describe any change in an object's velocity vector, which can be either a change in its magnitude or a change in its direction.

An important application of the vector addition of forces is the use of Newton's first law to analyze mechanical systems.

Homework Problems

1 ✓. A dinosaur fossil is slowly moving down the slope of a glacier under the influence of wind, rain and gravity. At the same time, the glacier is moving relative to the continent underneath. The dashed lines represent the directions but not the magnitudes of the velocities. Pick a scale, and use graphical addition of vectors to find the magnitude and the direction of the fossil's velocity relative to the continent. You will need a ruler and protractor.

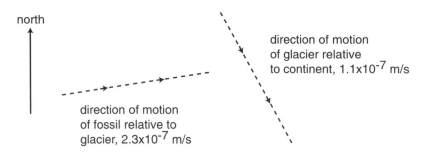

2. Is it possible for a helicopter to have an acceleration due east and a velocity due west? If so, what would be going on? If not, why not?

3 ✓. A bird is initially flying horizontally east at 21.1 m/s, but one second later it has changed direction so that it is flying horizontally and 7° north of east, at the same speed. What are the magnitude and direction of its acceleration vector during that one second time interval? (Assume its acceleration was roughly constant.)

4. A person of mass M stands in the middle of a tightrope, which is fixed at the ends to two buildings separated by a horizontal distance L. The rope sags in the middle, stretching and lengthening the rope slightly. (a✓) If the tightrope walker wants the rope to sag vertically by no more than a height h, find the minimum tension, T, that the rope must be able to withstand without breaking, in terms of h, g, M, and L. (b) Based on your equation, explain why it is not possible to get $h=0$, and give a physical interpretation.

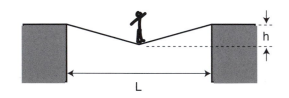

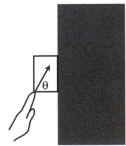

Problem 5.

5 ★✓. Your hand presses a block of mass m against a wall with a force F_H acting at an angle θ. Find the minimum and maximum possible values of $|F_H|$ that can keep the block stationary, in terms of m, g, θ, and μ_s, the coefficient of static friction between the block and the wall.

S A solution is given in the back of the book.   ★ A difficult problem.
✓ A computerized answer check is available.   ∫ A problem that requires calculus.

6✓. A skier of mass m is coasting down a slope inclined at an angle θ compared to horizontal. Assume for simplicity that the treatment of kinetic friction given in chapter 5 is appropriate here, although a soft and wet surface actually behaves a little differently. The coefficient of kinetic friction acting between the skis and the snow is μ_k, and in addition the skier experiences an air friction force of magnitude bv^2, where b is a constant. (a) Find the maximum speed that the skier will attain, in terms of the variables m, θ, μ_k, and b. (b) For angles below a certain minimum angle θ_{min}, the equation gives a result that is not mathematically meaningful. Find an equation for θ_{min}, and give a physical explanation of what is happening for $\theta < \theta_{min}$.

7 ∫. A gun is aimed horizontally to the west, and fired at $t=0$. The bullet's position vector as a function of time is $r = b\hat{x} + ct\hat{y} + dt^2\hat{z}$, where b, c, and d are constants. (a) What units would b, c, and d need to have for the equation to make sense? (b) Find the bullet's velocity and acceleration as functions of time. (c) Give physical interpretations of b, c, d, $\hat{x}$, $\hat{y}$, and $\hat{z}$.

8 S. Annie Oakley, riding north on horseback at 30 mi/hr, shoots her rifle, aiming horizontally and to the northeast. The muzzle speed of the rifle is 140 mi/hr. When the bullet hits a defenseless fuzzy animal, what is its speed of impact? Neglect air resistance, and ignore the vertical motion of the bullet.

9 S. A cargo plane has taken off from a tiny airstrip in the Andes, and is climbing at constant speed, at an angle of $\theta=17°$ with respect to horizontal. Its engines supply a thrust of $F_{thrust}=200$ kN, and the lift from its wings is $F_{lift}=654$ kN. Assume that air resistance (drag) is negligible, so the only forces acting are thrust, lift, and weight. What is its mass, in kg?

10 S. A wagon is being pulled at constant speed up a slope θ by a rope that makes an angle φ with the vertical. (a) Assuming negligible friction, show that the tension in the rope is given by the equation

$$F_T = \frac{\sin \theta}{\sin (\theta + \varphi)} F_W \quad ,$$

where F_W is the weight force acting on the wagon. (b) Interpret this equation in the special cases of $\varphi=0$ and $\varphi=180°-\theta$.

11 S. The angle of repose is the maximum slope on which an object will not slide. On airless, geologically inert bodies like the moon or an asteroid, the only thing that determines whether dust or rubble will stay on a slope is whether the slope is less steep than the angle of repose. (a) Find an equation for the angle of repose, deciding for yourself what are the relevant variables. (b) On an asteroid, where g can be thousands of times lower than on Earth, would rubble be able to lie at a steeper angle of repose?

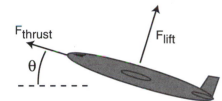

F_{thrust} F_{lift}

θ

Problem 9.

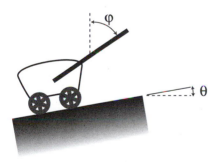

φ

θ

Problem 10.

9 Circular Motion

9.1 Conceptual Framework for Circular Motion

I now live fifteen minutes from Disneyland, so my friends and family in my native Northern California think it's a little strange that I've never visited the Magic Kingdom again since a childhood trip to the south. The truth is that for me as a preschooler, Disneyland was not the Happiest Place on Earth. My mother took me on a ride in which little cars shaped like rocket ships circled rapidly around a central pillar. I knew I was going to die. There was a force trying to throw me outward, and the safety features of the ride would surely have been inadequate if I hadn't screamed the whole time to make sure Mom would hold on to me. Afterward, she seemed surprisingly indifferent to the extreme danger we had experienced.

Circular motion does not produce an outward force

My younger self's understanding of circular motion was partly right and partly wrong. I was wrong in believing that there was a force pulling me outward, away from the center of the circle. The easiest way to understand this is to bring back the parable of the bowling ball in the pickup truck from chapter 4. As the truck makes a left turn, the driver looks in the rearview mirror and thinks that some mysterious force is pulling the ball outward, but the truck is accelerating, so the driver's frame of reference is not an inertial frame. Newton's laws are violated in a noninertial frame, so the ball appears to accelerate without any actual force acting on it. Because we are used to inertial frames, in which accelerations are caused by forces, the ball's acceleration creates a vivid illusion that there must be an outward force.

In an inertial frame everything makes more sense. The ball has no force on it, and goes straight as required by Newton's first law. The truck has a force on it from the asphalt, and responds to it by accelerating (changing the direction of its velocity vector) as Newton's second law says it should.

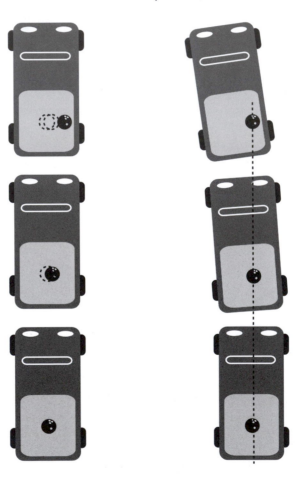

(a) In the turning truck's frame of reference, the ball appears to violate Newton's laws, displaying a sideways acceleration that is not the result of a force-interaction with any other object.

(b) In an inertial frame of reference, such as the frame fixed to the earth's surface, the ball obeys Newton's first law. No forces are acting on it, and it continues moving in a straight line. It is the truck that is participating in an interaction with the asphalt, the truck that accelerates as it should according to Newton's second law.

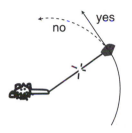

(a) An overhead view of a person swinging a rock on a rope. A force from the string is required to make the rock's velocity vector keep changing direction.

yes
no

(b) If the string breaks, the rock will follow Newton's first law and go straight instead of continuing around the circle.

Circular motion does not persist without a force

I was correct about one thing, however. To make me curve around with the car, I really did need some force such as a force from my mother, friction from the seat, or a normal force from the side of the car. (In fact, all three forces were probably adding together.) One of the reasons why Galileo failed to refine the principle of inertia into a quantitative statement like Newton's first law is that he was not sure whether motion without a force would naturally be circular or linear. In fact, the most impressive examples he knew of the persistence of motion were mostly circular: the spinning of a top or the rotation of the earth, for example. Newton realized that in examples such as these, there really were forces at work. Atoms on the surface of the top are prevented from flying off straight by the ordinary force that keeps atoms stuck together in solid matter. The earth is nearly all liquid, but gravitational forces pull all its parts inward.

Uniform and nonuniform circular motion

Circular motion always involves a change in the direction of the velocity vector, but it is also possible for the magnitude of the velocity to change at the same time. Circular motion is referred to as *uniform* if $|v|$ is constant, and *nonuniform* if it is changing.

Your speedometer tells you the magnitude of your car's velocity vector, so when you go around a curve while keeping your speedometer needle steady, you are executing uniform circular motion. If your speedometer reading is changing as you turn, your circular motion is nonuniform. Uniform circular motion is simpler to analyze mathematically, so we will attack it first and then pass to the nonuniform case.

Self-Check

Which of these are examples of uniform circular motion and which are nonuniform?
(a) the clothes in a clothes dryer (assuming they remain against the inside of the drum, even at the top)
(b) a rock on the end of a string being whirled in a vertical circle

(a) Uniform. They have the same motion as the drum itself, which is rotating as one solid piece. No part of the drum can be rotating at a different speed from any other part. (b) Nonuniform. Gravity speeds it up on the way down and slows it down on the way up.

Only an inward force is required for uniform circular motion.

The figures on the previous page showed the string pulling in straight along a radius of the circle, but many people believe that when they are doing this they must be "leading" the rock a little to keep it moving along. That is, they believe that the force required to produce uniform circular motion is not directly inward but at a slight angle to the radius of the circle. This intuition is incorrect, which you can easily verify for yourself now if you have some string handy. It is only while you are getting the object going that your force needs to be at an angle to the radius. During this initial period of speeding up, the motion is not uniform. Once you settle down into uniform circular motion, you only apply an inward force.

If you have not done the experiment for yourself, here is a theoretical argument to convince you of this fact. We have discussed in chapter 6 the principle that forces have no perpendicular effects. To keep the rock from speeding up or slowing down, we only need to make sure that our force is perpendicular to its direction of motion. We are then guaranteed that its forward motion will remain unaffected: our force can have no perpendicular effect, and there is no other force acting on the rock which could slow it down. The rock requires no forward force to maintain its forward motion, any more than a projectile needs a horizontal force to "help it over the top" of its arc.

Why, then, does a car driving in circles in a parking lot stop executing uniform circular motion if you take your foot off the gas? The source of confusion here is that Newton's laws predict an object's motion based on the *total* force acting on it. A car driving in circles has three forces on it

(1) an inward force from the asphalt, controlled with the steering wheel;
(2) a forward force from the asphalt, controlled with the gas pedal; and
(3) backward forces from air resistance and rolling resistance.

You need to make sure there is a forward force on the car so that the backward forces will be exactly canceled out, creating a vector sum that points directly inward.

In uniform circular motion, the acceleration vector is inward

Since experiments show that the force vector points directly inward, Newton's second law implies that the acceleration vector points inward as well. This fact can also be proven on purely kinematical grounds, and we will do so in the next section.

To make the brick go in a circle, I had to exert an inward force on the rope.

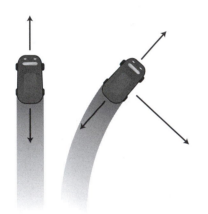

When a car is going straight at constant speed, the forward and backward forces on it are canceling out, producing a total force of zero. When it moves in a circle at constant speed, there are three forces on it, but the forward and backward forces cancel out, so the vector sum is an inward force.

A series of three hammer taps makes the rolling ball trace a triangle, seven hammers a heptagon. If the number of hammers was large enough, the ball would essentially be experiencing a steady inward force, and it would go in a circle. In no case is any forward force necessary.

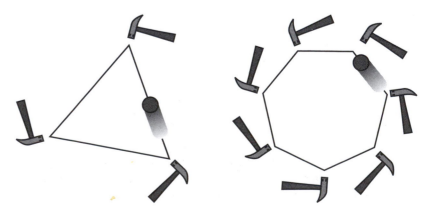

Discussion Questions

A. In the game of crack the whip, a line of people stand holding hands, and then they start sweeping out a circle. One person is at the center, and rotates without changing location. At the opposite end is the person who is running the fastest, in a wide circle. In this game, someone always ends up losing their grip and flying off. Suppose the person on the end loses her grip. What path does she follow as she goes flying off? (Assume she is going so fast that she is really just trying to put one foot in front of the other fast enough to keep from falling; she is not able to get any significant horizontal force between her feet and the ground.)

B. Suppose the person on the outside is still holding on, but feels that she may loose her grip at any moment. What force or forces are acting on her, and in what directions are they? (We are not interested in the vertical forces, which are the earth's gravitational force pulling down, and the ground's normal force pushing up.)

C. Suppose the person on the outside is still holding on, but feels that she may loose her grip at any moment. What is wrong with the following analysis of the situation? "The person whose hand she's holding exerts an inward force on her, and because of Newton's third law, there's an equal and opposite force acting outward. That outward force is the one she feels throwing her outward, and the outward force is what might make her go flying off, if it's strong enough."

D. If the only force felt by the person on the outside is an inward force, why doesn't she go straight in?

E. In the amusement park ride shown in the figure, the cylinder spins faster and faster until the customer can pick her feet up off the floor without falling. In the old Coney Island version of the ride, the floor actually dropped out like a trap door, showing the ocean below. (There is also a version in which the whole thing tilts up diagonally, but we're discussing the version that stays flat.) If there is no outward force acting on her, why does she stick to the wall? Analyze all the forces on her.

F. What is an example of circular motion where the inward force is a normal force? What is an example of circular motion where the inward force is friction? What is an example of circular motion where the inward force is the sum of more than one force?

G. Does the acceleration vector always change continuously in circular motion? The velocity vector?

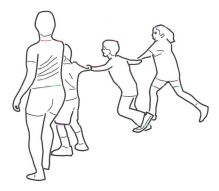

Discussion questions A-D.

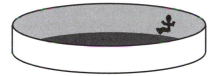

Discussion question E.

9.2 Uniform Circular Motion

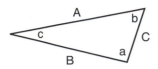

The law of sines:
$A/\sin a = B/\sin b = C/\sin c$

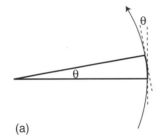

(a)

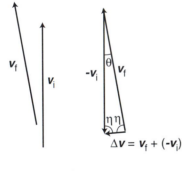

(b)   (c)

In this section I derive a simple and very useful equation for the magnitude of the acceleration of an object undergoing constant acceleration. The law of sines is involved, so I've recapped it on the left.

The derivation is brief, but the method requires some explanation and justification. The idea is to calculate a Δv vector describing the change in the velocity vector as the object passes through an angle θ. We then calculate the acceleration, $a=\Delta v/\Delta t$. The astute reader will recall, however, that this equation is only valid for motion with constant acceleration. Although the magnitude of the acceleration is constant for uniform circular motion, the acceleration vector changes its direction, so it is not a constant vector, and the equation $a=\Delta v/\Delta t$ does not apply. The justification for using it is that we will then examine its behavior when we make the time interval very short, which means making the angle θ very small. For smaller and smaller time intervals, the $\Delta v/\Delta t$ expression becomes a better and better approximation, so that the final result of the derivation is exact.

In figure (a), the object sweeps out an angle θ. Its direction of motion also twists around by an angle θ, from the vertical dashed line to the tilted one. Figure (b) shows the initial and final velocity vectors, which have equal magnitude, but directions differing by θ. In (c), the vectors have been reassembled in the proper orientation for vector subtraction. They form an isosceles triangle with interior angles θ, η, and η. (Eta, η, is my favorite Greek letter.) The law of sines gives

$$\frac{|\Delta v|}{\sin \theta} = \frac{|v|}{\sin \eta} \quad .$$

This tells us the magnitude of Δv, which is one of the two ingredients we need for calculating the magnitude of $a=\Delta v/\Delta t$. The other ingredient is Δt. The time required for the object to move through the angle θ is

$$\Delta t = \frac{\text{length of arc}}{|v|} \quad .$$

Now if we measure our angles in radians we can use the definition of radian measure, which is (angle)=(length of arc)/(radius), giving $\Delta t=\theta r/|v|$. Combining this with the first expression involving $|\Delta v|$ gives

$$|a| = |\Delta v|/\Delta t$$

$$= \frac{|v|^2}{r} \cdot \frac{\sin \theta}{\theta} \cdot \frac{1}{\sin \eta} \quad .$$

When θ becomes very small, the small-angle approximation $\sin \theta \approx \theta$ applies, and also η becomes close to 90°, so $\sin \eta \approx 1$, and we have an equation for $|a|$:

$$|a| = \frac{|v|^2}{r} \qquad \text{[uniform circular motion]} \quad .$$

Example: force required to turn on a bike

Question: A bicyclist is making a turn along an arc of a circle with radius 20 m, at a speed of 5 m/s. If the combined mass of the cyclist plus the bike is 60 kg, how great a static friction force must the road be able to exert on the tires?

Solution: Taking the magnitudes of both sides of Newton's second law gives

$$|\mathbf{F}| = |m\mathbf{a}|$$
$$= m|\mathbf{a}| \quad .$$

Substituting $|\mathbf{a}| = |\mathbf{v}|^2/r$ gives

$$|\mathbf{F}| = m|\mathbf{v}|^2/r$$
$$\approx 80 \text{ N}$$

(rounded off to one sig fig).

Example: Don't hug the center line on a curve!

Question: You're driving on a mountain road with a steep drop on your right. When making a left turn, is it safer to hug the center line or to stay closer to the outside of the road?

Solution: You want whichever choice involves the least acceleration, because that will require the least force and entail the least risk of exceeding the maximum force of static friction. Assuming the curve is an arc of a circle and your speed is constant, your car is performing uniform circular motion, with $|\mathbf{a}| = |\mathbf{v}|^2/r$. The dependence on the square of the speed shows that driving slowly is the main safety measure you can take, but for any given speed you also want to have the largest possible value of r. Even though your instinct is to keep away from that scary precipice, you are actually less likely to skid if you keep toward the outside, because then you are describing a larger circle.

Example: acceleration related to radius and period of rotation

Question: How can the equation for the acceleration in uniform circular motion be rewritten in terms of the radius of the circle and the *period*, T, of the motion, i.e. the time required to go around once?

Solution: The period can be related to the speed as follows:

$$|\mathbf{v}| = \frac{\text{circumference}}{T}$$
$$= 2\pi r/T \quad .$$

Substituting into the equation $|\mathbf{a}| = |\mathbf{v}|^2/r$ gives

$$|\mathbf{a}| = \frac{4\pi^2 r}{T^2} \quad .$$

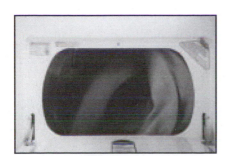

Example: a clothes dryer

Question: My clothes dryer has a drum with an inside radius of 35 cm, and it spins at 48 revolutions per minute. What is the acceleration of the clothes inside?

Solution: We can solve this by finding the period and plugging in to the result of the previous example. If it makes 48 revolutions in one minute, then the period is 1/48 of a minute, or 1.25 s. To get an acceleration in mks units, we must convert the radius to 0.35 m. Plugging in, the result is 8.8 m/s².

Example: more about clothes dryers!

Question: In a discussion question in the previous section, we made the assumption that the clothes remain against the inside of the drum as they go over the top. In light of the previous example, is this a correct assumption?

Solution: No. We know that there must be some minimum speed at which the motor can run that will result in the clothes just barely staying against the inside of the drum as they go over the top. If the clothes dryer ran at just this minimum speed, then there would be no normal force on the clothes at the top: they would be on the verge of losing contact. The only force acting on them at the top would be the force of gravity, which would give them an acceleration of g=9.8 m/s^2. The actual dryer must be running slower than this minimum speed, because it produces an acceleration of only 8.8 m/s^2. My theory is that this is done intentionally, to make the clothes mix and tumble.

Discussion Question

A. A certain amount of force is needed to provide the acceleration of circular motion. What if were are exerting a force perpendicular to the direction of motion in an attempt to make an object trace a circle of radius r, but the force isn't as big as $m|v|^2/r$?

B. Suppose a rotating space station is built that gives its occupants the illusion of ordinary gravity. What happens when a person in the station lets go of a ball? What happens when she throws a ball straight "up" in the air (i.e. towards the center)?

An artist's conception of a rotating space colony in the form of a giant wheel. A person living in this noninertial frame of reference has an illusion of a force pulling her outward, toward the deck, for the same reason that a person in the pickup truck has the illusion of a force pulling the bowling ball. By adjusting the speed of rotation, the designers can make an acceleration $|v|^2/r$ equal to the usual acceleration of gravity on earth. On earth, your acceleration standing on the ground is zero, and a falling rock heads for your feet with an acceleration of 9.8 m/s^2. A person standing on the deck of the space colony has an *upward* acceleration of 9.8 m/s^2, and when she lets go of a rock, her feet head *up* at the nonaccelerating rock. To her, it seems the same as true gravity.
Art by NASA.

9.3 Nonuniform Circular Motion

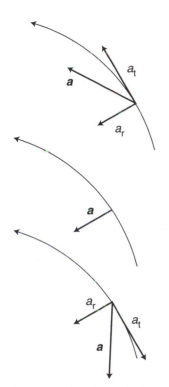

An object moving in a circle may speed up (top), keep the magnitude of its velocity vector constant (middle), or slow down (bottom).

What about nonuniform circular motion? Although so far we have been discussing components of vectors along fixed x and y axes, it now becomes convenient to discuss components of the acceleration vector along the radial line (in-out) and the tangential line (along the direction of motion). For nonuniform circular motion, the radial component of the acceleration obeys the same equation as for uniform circular motion,

$$a_r = |v|^2/r \quad ,$$

but the acceleration vector also has a tangential component,

$$a_t = \text{slope of the graph of } |v| \text{ versus } t \ .$$

The latter quantity has a simple interpretation. If you are going around a curve in your car, and the speedometer needle is moving, the tangential component of the acceleration vector is simply what you would have thought the acceleration was if you saw the speedometer and didn't know you were going around a curve.

> Example: Slow down before a turn, not during it.
> **Question**: When you're making a turn in your car and you're afraid you may skid, isn't it a good idea to slow down?
> **Solution**: If the turn is an arc of a circle, and you've already completed part of the turn at constant speed without skidding, then the road and tires are apparently capable of enough static friction to supply an acceleration of $|v|^2/r$. There is no reason why you would skid out now if you haven't already. If you get nervous and brake, however, then you need to have a tangential acceleration component in addition to the radial component you were already able to produce successfully. This would require an acceleration vector with a greater magnitude, which in turn would require a larger force. Static friction might not be able to supply that much force, and you might skid out. As in the previous example on a similar topic, the safe thing to do is to approach the turn at a comfortably low speed.

Summary

Selected Vocabulary

uniform circular motion .... circular motion in which the magnitude of the velocity vector remains constant

nonuniform circular motion ... circular motion in which the magnitude of the velocity vector changes

radial .... parallel to the radius of a circle; the in-out direction

tangential .... tangent to the circle, perpendicular to the radial direction

Notation

a_r .... radial acceleration; the component of the acceleration vector along the in-out direction

a_t .... tangential acceleration; the component of the acceleration vector tangent to the circle

Summary

If an object is to have circular motion, a force must be exerted on it toward the center of the circle. There is no outward force on the object; the illusion of an outward force comes from our experiences in which our point of view was rotating, so that we were viewing things in a noninertial frame.

An object undergoing uniform circular motion has an inward acceleration vector of magnitude

$$|\mathbf{a}| = \frac{|v|^2}{r} \quad .$$

In nonuniform circular motion, the radial and tangential components of the acceleration vector are

$$a_r = |\mathbf{v}|^2/r$$

$$a_t = \text{slope of the graph of } |\mathbf{v}| \text{ versus } t \quad .$$

Homework Problems

1. When you're done using an electric mixer, you can get most of the batter off of the beaters by lifting them out of the batter with the motor running at a high enough speed. Let's imagine, to make things easier to visualize, that we instead have a piece of tape stuck to one of the beaters. (a) Explain why static friction has no effect on whether or not the tape flies off. (b) Suppose you find that the tape doesn't fly off when the motor is on a low speed, but speeding it up does cause it to fly off. Why would the greater speed change things?

2. Show that the expression $|v|^2/r$ has the units of acceleration.

3 ✓. A plane is flown in a loop-the-loop of radius 1.00 km. The plane starts out flying upside-down, straight and level, then begins curving up along the circular loop, and is right-side up when it reaches the top . (The plane may slow down somewhat on the way up.) How fast must the plane be going at the top if the pilot is to experience no force from the seat or the seatbelt while at the top of the loop?

4 ∫. In this problem, you'll derive the equation $|a|=|v|^2/r$ using calculus. Instead of comparing velocities at two points in the particle's motion and then taking a limit where the points are close together, you'll just take derivatives. The particle's position vector is $r=(r \cos \theta)\hat{x} + (r \sin \theta)\hat{y}$, where $\hat{x}$ and $\hat{y}$ are the unit vectors along the x and y axes. By the definition of radians, the distance traveled since $t=0$ is $r\theta$, so if the particle is traveling at constant speed $v=|v|$, we have $v=r\theta/t$. (a) Eliminate θ to get the particle's position vector as a function of time. (b) Find the particle's acceleration vector. (c) Show that the magnitude of the acceleration vector equals v^2/r.

5 S. Three cyclists in a race are rounding a semicircular curve. At the moment depicted, cyclist A is using her brakes to apply a force of 375 N to her bike. Cyclist B is coasting. Cyclist C is pedaling, resulting in a force of 375 N on her bike. Each cyclist, with her bike, has a mass of 75 kg. At the instant shown, the instantaneous speed of all three cyclists is 10 m/s. On the diagram, draw each cyclist's acceleration vector with its tail on top of her present position, indicating the directions and lengths reasonably accurately. Indicate approximately the consistent scale you are using for all three acceleration vectors. Extreme precision is not necessary as long as the directions are approximately right, and lengths of vectors that should be equal appear roughly equal, etc. Assume all three cyclists are traveling along the road all the time, not wandering across their lane or wiping out and going off the road.

20 m

direction
of travel

C

B

A

Problem 5.

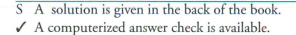

6 S★. The amusement park ride shown in the figure consists of a cylindrical room that rotates about its vertical axis. When the rotation is fast enough, a person against the wall can pick his or her feet up off the floor and remain "stuck" to the wall without falling.

(a) Suppose the rotation results in the person having a speed v. The radius of the cylinder is r, the person's mass is m, the downward acceleration of gravity is g, and the coefficient of static friction between the person and the wall is μ_s. Find an equation for the speed, v, required, in terms of the other variables. (You will find that one of the variables cancels out.)

(b) Now suppose two people are riding the ride. Huy is wearing denim, and Gina is wearing polyester, so Huy's coefficient of static friction is three times greater. The ride starts from rest, and as it begins rotating faster and faster, Gina must wait longer before being able to lift her feet without sliding to the floor. Based on your equation from part a, how many times greater must the speed be before Gina can lift her feet without sliding down?

7 S. An engineer is designing a curved off-ramp for a freeway. Since the off-ramp is curved, she wants to bank it to make it less likely that motorists going too fast will wipe out. If the radius of the curve is r, how great should the banking angle, θ, be so that for a car going at a speed v, no static friction force whatsoever is required to allow the car to make the curve? State your answer in terms of v, r, and g, and show that the mass of the car is irrelevant.

8 ✓. Lionel brand toy trains come with sections of track in standard lengths and shapes. For circular arcs, the most commonly used sections have diameters of 662 and 1067 mm at the inside of the outer rail. The maximum speed at which a train can take the broader curve without flying off the tracks is 0.95 m/s. At what speed must the train be operated to avoid derailing on the tighter curve?

9. The figure shows a ball on the end of a string of length L attached to a vertical rod which is spun about its vertical axis by a motor. The period (time for one rotation) is P.

(a) Analyze the forces in which the ball participates.

(b✓) Find how the angle θ depends on P, g, and L. [Hints: (1) Write down Newton's second law for the vertical and horizontal components of force and acceleration. This gives two equations, which can be solved for the two unknowns, θ and the tension in the string. (2) If you introduce variables like v and r, relate them to the variables your solution is supposed to contain, and eliminate them.]

(c) What happens mathematically to your solution if the motor is run very slowly (very large values of P)? Physically, what do you think would actually happen in this case?

Problem 6.

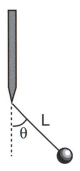

Problem 7.

Problem 9.

Problem 10.

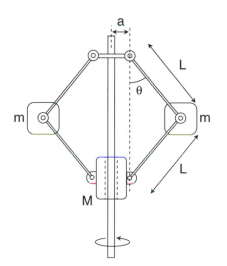

Problem 11.

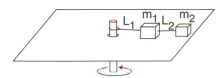

Problem 12.

10. Psychology professor R.O. Dent requests funding for an experiment on compulsive thrill-seeking behavior in hamsters, in which the subject is to be attached to the end of a spring and whirled around in a horizontal circle. The spring has equilibrium length b, and obeys Hooke's law with spring constant k. It is stiff enough to keep from bending significantly under the hamster's weight.

(a✔) Calculate the length of the spring when it is undergoing steady circular motion in which one rotation takes a time T. Express your result in terms of k, b, and T.

(b) The ethics committee somehow fails to veto the experiment, but the safety committee expresses concern. Why? Does your equation do anything unusual, or even spectacular, for any particular value of T? What do you think is the physical significance of this mathematical behavior?

11★. The figure shows an old-fashioned device called a flyball governor, used for keeping an engine running at the correct speed. The whole thing rotates about the vertical shaft, and the mass M is free to slide up and down. This mass would have a connection (not shown) to a valve that controlled the engine. If, for instance, the engine ran too fast, the mass would rise, causing the engine to slow back down.

(a) Show that in the special case of $a=0$, the angle θ is given by

$$\theta = \cos^{-1}\left(\frac{g(m+M)P^2}{4\pi^2 mL}\right) \quad ,$$

where P is the period of rotation (time required for one complete rotation).

(b) There is no closed-form solution for θ in the general case where a is not zero. However, explain how the undesirable low-speed behavior of the $a=0$ device would be improved by making a nonzero.

[Based on an example by J.P. den Hartog.]

12✔. The figure shows two blocks of masses m_1 and m_2 sliding in circles on a frictionless table. Find the tension in the strings if the period of rotation (time required for one complete rotation) is P.

13. The acceleration of an object in uniform circular motion can be given either by $|a|=|v|^2/r$ or, equivalently, by $|a|=4\pi^2 r/T^2$, where T is the time required for one cycle. (The latter expression comes simply from substituting $|v|=$distance/time=circumference/$T=2\pi r/T$ into the first expression.) Person A says based on the first equation that the acceleration in circular motion is greater when the circle is smaller. Person B, arguing from the second equation, says that the acceleration is smaller when the circle is smaller. Rewrite the two statements so that they are less misleading, eliminating the supposed paradox. [Based on a problem by Arnold Arons.]

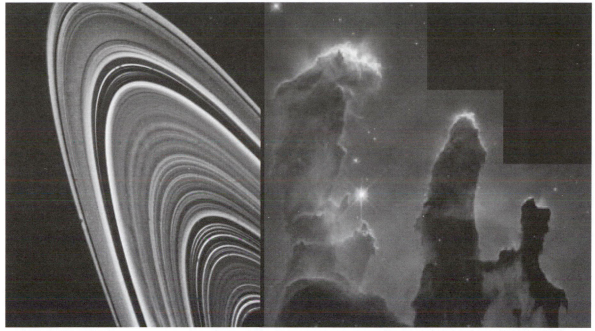

Gravity is the only really important force on the cosmic scale. Left: a false-color image of saturn's rings, composed of innumerable tiny ice particles orbiting in circles under the influence of saturn's gravity. Right: A stellar nursery, the Eagle Nebula. Each pillar of hydrogen gas is about as tall as the diameter of our entire solar system. The hydrogen molecules all attract each other through gravitational forces, resulting in the formation of clumps that contract to form new stars.

10 Gravity

Cruise your radio dial today and try to find any popular song that would have been imaginable without Louis Armstrong. By introducing solo improvisation into jazz, Armstrong took apart the jigsaw puzzle of popular music and fit the pieces back together in a different way. In the same way, Newton reassembled our view of the universe. Consider the titles of some recent physics books written for the general reader: **The God Particle**, **Dreams of a Final Theory**. When the subatomic particle called the neutrino was recently proven for the first time to have mass, specialists in cosmology began discussing seriously what effect this would have on calculations of the ultimate fate of the universe: would the neutrinos' mass cause enough extra gravitational attraction to make the universe eventually stop expanding and fall back together? Without Newton, such attempts at universal understanding would not merely have seemed a little pretentious, they simply would not have occurred to anyone.

This chapter is about Newton's theory of gravity, which he used to explain the motion of the planets as they orbited the sun. Whereas this book has concentrated on Newton's laws of motion, leaving gravity as a dessert, Newton tosses off the laws of motion in the first 20 pages of the **Principia Mathematica** and then spends the next 130 discussing the motion of the planets. Clearly he saw this as the crucial scientific focus of his work. Why? Because in it he showed that the same laws of motion applied to the heavens as to the earth, and that the gravitational force that made an apple fall was the same as the force that kept the earth's motion from carrying it away from the sun. What was radical about Newton was not his laws of motion but his concept of a universal science of physics.

10.1 Kepler's Laws

Tycho Brahe made his name as an astronomer by showing that the bright new star, today called a supernova, that appeared in the skies in 1572 was far beyond the Earth's atmosphere. This, along with Galileo's discovery of sunspots, showed that contrary to Aristotle, the heavens were not perfect and unchanging. Brahe's fame as an astronomer brought him patronage from King Frederick II, allowing him to carry out his historic high-precision measurements of the planets' motions. A contradictory character, Brahe enjoyed lecturing other nobles about the evils of dueling, but had lost his own nose in a youthful duel and had it replaced with a prosthesis made of an alloy of gold and silver. Willing to endure scandal in order to marry a peasant, he nevertheless used the feudal powers given to him by the king to impose harsh forced labor on the inhabitants of his parishes. The result of their work, an Italian-style palace with an observatory on top, surely ranks as one of the most luxurious science labs ever built. When the king died and his son reduced Brahe's privileges, Brahe left in a huff for a new position in Prague, taking his data with him. He died of a ruptured bladder after falling from a wagon on the way home from a party — in those days, it was considered rude to leave the dinner table to relieve oneself.

Newton wouldn't have been able to figure out *why* the planets move the way they do if it hadn't been for the astronomer Tycho Brahe (1546-1601) and his protege Johannes Kepler (1571-1630), who together came up with the first simple and accurate description of *how* the planets actually do move. The difficulty of their task is suggested by the figure below, which shows how the relatively simple orbital motions of the earth and Mars combine so that as seen from earth Mars appears to be staggering in loops like a drunken sailor.

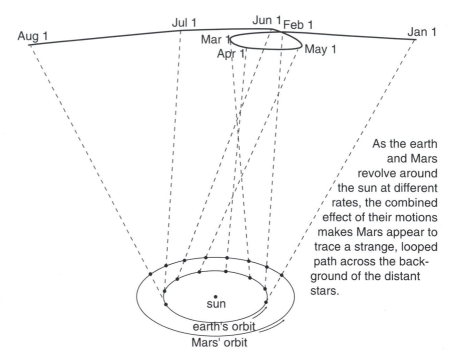

As the earth and Mars revolve around the sun at different rates, the combined effect of their motions makes Mars appear to trace a strange, looped path across the background of the distant stars.

Brahe, the last of the great naked-eye astronomers, collected extensive data on the motions of the planets over a period of many years, taking the giant step from the previous observations' accuracy of about 10 seconds of arc (10/60 of a degree) to an unprecedented 1 second. The quality of his work is all the more remarkable considering that his observatory consisted of four giant brass protractors mounted upright in his castle in Denmark. Four different observers would simultaneously measure the position of a planet in order to check for mistakes and reduce random errors.

With Brahe's death, it fell to his former assistant Kepler to try to make some sense out of the volumes of data. Kepler, in contradiction to his late boss, had formed a prejudice, a correct one as it turned out, in favor of the theory that the earth and planets revolved around the sun, rather than the earth staying fixed and everything rotating about it. Although motion is relative, it is not just a matter of opinion what circles what. The earth's rotation and revolution about the sun make it a noninertial reference frame, which causes detectable violations of Newton's laws when one attempts to describe sufficiently precise experiments in the earth-fixed frame. Although such direct experiments were not carried out until the 19th century, what

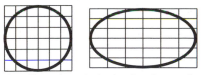

An ellipse is a circle that has been distorted by shrinking and stretching along perpendicular axes.

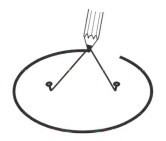

An ellipse can be constructed by tying a string to two pins and drawing like this with the pencil stretching the string taut. Each pin constitutes one focus of the ellipse.

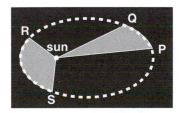

If the time interval taken by the planet to move from P to Q is equal to the time interval from R to S, then according to Kepler's equal-area law, the two shaded areas are equal. The planet is moving faster during interval RS than it did during PQ, which Newton later determined was due to the sun's gravitational force accelerating it. The equal-area law predicts exactly how much it will speed up.

convinced everyone of the sun-centered system in the 17th century was that Kepler was able to come up with a surprisingly simple set of mathematical and geometrical rules for describing the planets' motion using the sun-centered assumption. After 900 pages of calculations and many false starts and dead-end ideas, Kepler finally synthesized the data into the following three laws:

Kepler's elliptical orbit law: The planets orbit the sun in elliptical orbits with the sun at one focus.

Kepler's equal-area law: The line connecting a planet to the sun sweeps out equal areas in equal amounts of time.

Kepler's law of periods: The time required for a planet to orbit the sun, called its period, is proportional to the long axis of the ellipse raised to the 3/2 power. The constant of proportionality is the same for all the planets.

Although the planets' orbits are ellipses rather than circles, most are very close to being circular. The earth's orbit, for instance, is only flattened by 1.7% relative to a circle. In the special case of a planet in a circular orbit, the two foci (plural of "focus") coincide at the center of the circle, and Kepler's elliptical orbit law thus says that the circle is centered on the sun. The equal-area law implies that a planet in a circular orbit moves around the sun with constant speed. For a circular orbit, the law of periods then amounts to a statement that the time for one orbit is proportional to $r^{3/2}$, where r is the radius. If all the planets were moving in their orbits at the same speed, then the time for one orbit would simply depend on the circumference of the circle, so it would only be proportional to r to the first power. The more drastic dependence on $r^{3/2}$ means that the outer planets must be moving more slowly than the inner planets.

10.2   Newton's Law of Gravity

The sun's force on the planets obeys an inverse square law.

Kepler's laws were a beautifully simple explanation of what the planets did, but they didn't address why they moved as they did. Did the sun exert a force that pulled a planet toward the center of its orbit, or, as suggested by Descartes, were the planets circulating in a whirlpool of some unknown liquid? Kepler, working in the Aristotelian tradition, hypothesized not just an inward force exerted by the sun on the planet, but also a second force in the direction of motion to keep the planet from slowing down. Some speculated that the sun attracted the planets magnetically.

Once Newton had formulated his laws of motion and taught them to some of his friends, they began trying to connect them to Kepler's laws. It was clear now that an inward force would be needed to bend the planets' paths. This force was presumably an attraction between the sun and each planet. (Although the sun does accelerate in response to the attractions of the planets, its mass is so great that the effect had never been detected by the prenewtonian astronomers.) Since the outer planets were moving slowly along more gently curving paths than the inner planets, their accelerations were apparently less. This could be explained if the sun's force was determined by distance, becoming weaker for the farther planets. Physicists were also familiar with the noncontact forces of electricity and magnetism, and knew that they fell off rapidly with distance, so this made sense.

In the approximation of a circular orbit, the magnitude of the sun's force on the planet would have to be

$$F \;=\; ma \;=\; mv^2/r \;. \tag{1}$$

Now although this equation has the magnitude, v, of the velocity vector in it, what Newton expected was that there would be a more fundamental underlying equation for the force of the sun on a planet, and that that equation would involve the distance, r, from the sun to the object, but not the object's speed, v — motion doesn't make objects lighter or heavier.

Self-Check

 If eq. (1) really was generally applicable, what would happen to an object released at rest in some empty region of the solar system?

Equation (1) was thus a useful piece of information which could be related to the data on the planets simply because the planets happened to be going in nearly circular orbits, but Newton wanted to combine it with other equations and eliminate v algebraically in order to find a deeper truth.

To eliminate v, Newton used the equation

$$v \;=\; \frac{\text{circumference}}{T}$$
$$=\; 2\pi r/T \;. \tag{2}$$

Of course this equation would also only be valid for planets in nearly circular orbits. Plugging this into eq. (1) to eliminate v gives

$$F \;=\; \frac{4\pi^2 mr}{T^2} \;. \tag{3}$$

This unfortunately has the side-effect of bringing in the period, T, which we expect on similar physical grounds will not occur in the final answer. That's where the circular-orbit case, $T \propto r^{3/2}$, of Kepler's law of periods comes in. Using it to eliminate T gives a result that depends only on the mass of the planet and its distance from the sun:

$$F \;\propto\; m/r^2 \;. \qquad \text{[force of the sun on a planet of}$$
mass m at a distance r from the sun; same proportionality constant for all the planets]

(Since Kepler's law of periods is only a proportionality, the final result is a proportionality rather than an equation, and there is this no point in hanging on to the factor of $4\pi^2$.)

 It would just stay where it was. Plugging v=0 into eq. (1) would give F=0, so it would not accelerate from rest, and would never fall into the sun. No astronomer had ever observed an object that did that!

As an example, the "twin planets" Uranus and Neptune have nearly the same mass, but Neptune is about twice as far from the sun as Uranus, so the sun's gravitational force on Neptune is about four times smaller.

The forces between heavenly bodies are the same type of force as terrestrial gravity

OK, but what kind of force was it? It probably wasn't magnetic, since magnetic forces have nothing to do with mass. Then came Newton's great insight. Lying under an apple tree and looking up at the moon in the sky, he saw an apple fall. Might not the earth also attract the moon with the same kind of gravitational force? The moon orbits the earth in the same way that the planets orbit the sun, so maybe the earth's force on the falling apple, the earth's force on the moon, and the sun's force on a planet were all the same type of force.

There was an easy way to test this hypothesis numerically. If it was true, then we would expect the gravitational forces exerted by the earth to follow the same $F \propto m/r^2$ rule as the forces exerted by the sun, but with a different constant of proportionality appropriate to the earth's gravitational strength. The issue arises now of how to define the distance, r, between the earth and the apple. An apple in England is closer to some parts of the earth than to others, but suppose we take r to be the distance from the center of the earth to the apple, i.e. the radius of the earth. (The issue of how to measure r did not arise in the analysis of the planets' motions because the sun and planets are so small compared to the distances separating them.) Calling the proportionality constant k, we have

$$F_{\text{earth on apple}} = k\, m_{\text{apple}}\, / \, r_{\text{earth}}^{\,2}$$
$$F_{\text{earth on moon}} = k\, m_{\text{moon}}\, / \, d_{\text{earth-moon}}^{\,2}\,.$$

Newton's second law says $a = F/m$, so

$$a_{\text{apple}} = k\, / \, r_{\text{earth}}^{\,2}$$
$$a_{\text{moon}} = k\, / \, d_{\text{earth-moon}}^{\,2}\,.$$

The Greek astronomer Hipparchus had already found 2000 years before that the distance from the earth to the moon was about 60 times the radius of the earth, so if Newton's hypothesis was right, the acceleration of the moon would have to be $60^2 = 3600$ times less than the acceleration of the falling apple.

Applying $a = v^2/r$ to the acceleration of the moon yielded an acceleration that was indeed 3600 times smaller than 9.8 m/s², and Newton was convinced he had unlocked the secret of the mysterious force that kept the moon and planets in their orbits.

Newton's law of gravity

The proportionality $F \propto m/r^2$ for the gravitational force on an object of mass m only has a consistent proportionality constant for various objects if they are being acted on by the gravity of the same object. Clearly the sun's gravitational strength is far greater than the earth's, since the planets all orbit the sun and do not exhibit any very large accelerations caused by the earth (or by one another). What property of the sun gives it its great gravitational strength? Its great volume? Its great mass? Its great temperature? Newton reasoned that if the force was proportional to the mass of the object being

60

1

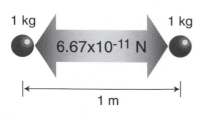

The gravitational attraction between two 1-kg masses separated by a distance of 1 m is 6.67x10⁻¹¹ N. Do not memorize this number!

acted on, then it would also make sense if the determining factor in the gravitational strength of the object exerting the force was its own mass. Assuming there were no other factors affecting the gravitational force, then the only other thing needed to make quantitative predictions of gravitational forces would be a proportionality constant. Newton called that proportionality constant G, and the complete form of the law of gravity he hypothesized was

$$F = Gm_1m_2/r^2 \ .$$ [gravitational force between objects of mass m_1 and m_2, separated by a distance r; r is not the radius of anything]

Newton conceived of gravity as an attraction between any two masses in the universe. The constant G tells us the how many newtons the attractive force is for two 1-kg masses separated by a distance of 1 m. The experimental determination of G in ordinary units (as opposed to the special, nonmetric, units used in astronomy) is described in section 10.5. This difficult measurement was not accomplished until long after Newton's death.

Example: The units of G
Question: What are the units of G?
Solution: Solving for G in Newton's law of gravity gives

$$G = \frac{Fr^2}{m_1m_2} \quad ,$$

so the units of G must be N · m²/ kg². Fully adorned with units, the value of G is 6.67x10⁻¹¹ N · m²/ kg².

Example: Newton's third law
Question: Is Newton's law of gravity consistent with Newton's third law?
Solution: The third law requires two things. First, m_1's force on m_2 should be the same as m_2's force on m_1. This works out, because the product m_1m_2 gives the same result if we interchange the labels 1 and 2. Second, the forces should be in opposite directions. This condition is also satisfied, because Newton's law of gravity refers to an attraction: each mass pulls the other toward itself.

Example: Pluto and Charon
Question: Pluto's moon Charon is unusually large considering Pluto's size, giving them the character of a double planet. Their masses are 1.25x10²² and 1.9x19²¹ kg, and their average distance from one another is 1.96x10⁴ km. What is the gravitational force between them?
Solution: If we want to use the value of G expressed in SI (meter-kilogram-second) units, we first have to convert the distance to 1.96x10⁷ m. The force is

$$\frac{\left(6.67\times10^{-11} \ \frac{N \cdot m^2}{kg^2}\right)\left(1.25 \times 10^{22} \ kg\right)\left(1.9 \times 10^{21} \ kg\right)}{\left(1.96 \times 10^7 \ m\right)^2}$$

$= 4.1 \times 10^{18}$ N

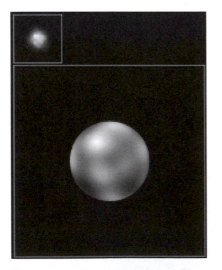

Computer-enhanced images of Pluto and Charon, taken by the Hubble Space Telescope.

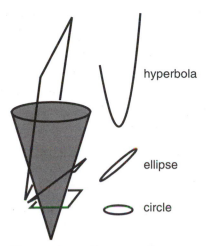

The conic sections are the curves made by cutting the surface of an infinite cone with a plane.

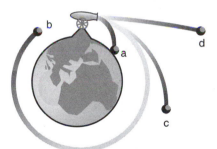

An imaginary cannon able to shoot cannonballs at very high speeds is placed on top of an imaginary, very tall mountain that reaches up above the atmosphere. Depending on the speed at which the ball is fired, it may end up in a tightly curved elliptical orbit, a, a circular orbit, b, a bigger elliptical orbit, c, or a nearly straight hyperbolic orbit, d.

The proportionality to $1/r^2$ in Newton's law of gravity was not entirely unexpected. Proportionalities to $1/r^2$ are found in many other phenomena in which some effect spreads out from a point. For instance, the intensity of the light from a candle is proportional to $1/r^2$, because at a distance r from the candle, the light has to be spread out over the surface of an imaginary sphere of area $4\pi r^2$. The same is true for the intensity of sound from a firecracker, or the intensity of gamma radiation emitted by the Chernobyl reactor. It's important, however, to realize that this is only an analogy. Force does not travel through space as sound or light does, and force is not a substance that can be spread thicker or thinner like butter on toast.

Although several of Newton's contemporaries had speculated that the force of gravity might be proportional to $1/r^2$, none of them, even the ones who had learned Newton's laws of motion, had had any luck proving that the resulting orbits would be ellipses, as Kepler had found empirically. Newton did succeed in proving that elliptical orbits would result from a $1/r^2$ force, but we postpone the proof until the end of the next volume of the textbook because it can be accomplished much more easily using the concepts of energy and angular momentum.

Newton also predicted that orbits in the shape of hyperbolas should be possible, and he was right. Some comets, for instance, orbit the sun in very elongated ellipses, but others pass through the solar system on hyperbolic paths, never to return. Just as the trajectory of a faster baseball pitch is flatter than that of a more slowly thrown ball, so the curvature of a planet's orbit depends on its speed. A spacecraft can be launched at relatively low speed, resulting in a circular orbit about the earth, or it can be launched at a higher speed, giving a more gently curved ellipse that reaches farther from the earth, or it can be launched at a very high speed which puts it in an even less curved hyperbolic orbit. As you go very far out on a hyperbola, it approaches a straight line, i.e. its curvature eventually becomes nearly zero.

Newton also was able to prove that Kepler's second law (sweeping out equal areas in equal time intervals) was a logical consequence of his law of gravity. Newton's version of the proof is moderately complicated, but the proof becomes trivial once you understand the concept of angular momentum, which will be covered later in the course. The proof will therefore be deferred until section 5.7 of book 2.

Self-Check

Which of Kepler's laws would it make sense to apply to hyperbolic orbits?

Discussion Questions

A. How could Newton find the speed of the moon to plug in to $a = v^2/r$?

B. Two projectiles of different mass shot out of guns on the surface of the earth at the same speed and angle will follow the same trajectories, assuming that air friction is negligible. (You can verify this by throwing two objects together from your hand and seeing if they separate or stay side by side.) What corresponding fact would be true for satellites of the earth having different masses?

C. What is wrong with the following statement? "A comet in an elliptical orbit speeds up as it approaches the sun, because the sun's force on it is increasing."

D. Why would it not make sense to expect the earth's gravitational force on a bowling ball to be inversely proportional to the square of the distance between their surfaces rather than their centers?

E. Does the earth accelerate as a result of the moon's gravitational force on it? Suppose two planets were bound to each other gravitationally the way the earth and moon are, but the two planets had equal masses. What would their motion be like?

F. Spacecraft normally operate by firing their engines only for a few minutes at a time, and an interplanetary probe will spend months or years on its way to its destination without thrust. Suppose a spacecraft is in a circular orbit around Mars, and it then briefly fires its engines in reverse, causing a sudden decrease in speed. What will this do to its orbit? What about a forward thrust?

10.3   Apparent Weightlessness

If you ask somebody at the bus stop why astronauts are weightless, you'll probably get one of the following two incorrect answers:

(1) They're weightless because they're so far from the earth.

(2) They're weightless because they're moving so fast.

The first answer is wrong, because the vast majority of astronauts never get more than a thousand miles from the earth's surface. The reduction in gravity caused by their altitude is significant, but not 100%. The second answer is wrong because Newton's law of gravity only depends on distance, not speed.

The correct answer is that astronauts in orbit around the earth are not really weightless at all. Their weightlessness is only apparent. If there was no gravitational force on the spaceship, it would obey Newton's first law and move off on a straight line, rather than orbiting the earth. Likewise, the astronauts inside the spaceship are in orbit just like the spaceship itself, with the earth's gravitational force continually twisting their velocity vectors around. The reason they appear to be weightless is that they are in the same orbit as the spaceship, so although the earth's gravity curves their trajectory down toward the deck, the deck drops out from under them at the same rate.

 The equal-area law makes equally good sense in the case of a hyperbolic orbit (and observations verify it). The elliptical orbit law had to be generalized by Newton to include hyperbolas. The law of periods doesn't make sense in the case of a hyperbolic orbit, because a hyperbola never closes back on itself, so the motion never repeats.

Apparent weightlessness can also be experienced on earth. Any time you jump up in the air, you experience the same kind of apparent weightlessness that the astronauts do. While in the air, you can lift your arms more easily than normal, because gravity does not make them fall any faster than the rest of your body, which is falling out from under them. The Russian air force now takes rich foreign tourists up in a big cargo plane and gives them the feeling of weightlessness for a short period of time while the plane is nose-down and dropping like a rock.

10.4 Vector Addition of Gravitational Forces

Pick a flower on earth and you move the farthest star.
Paul Dirac

When you stand on the ground, which part of the earth is pulling down on you with its gravitational force? Most people are tempted to say that the effect only comes from the part directly under you, since gravity always pulls straight down. Here are three observations that might help to change your mind:

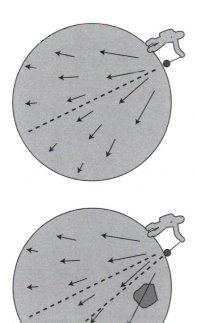

Gravity only appears to pull straight down because the near perfect symmetry of the earth makes the sideways components of the total force on an object cancel almost exactly. If the symmetry is broken, e.g. by a dense mineral deposit, the total force is a little off to the side.

- If you jump up in the air, gravity does not stop affecting you just because you are not touching the earth: gravity is a noncontact force. That means you are not immune from the gravity of distant parts of our planet just because you are not touching them.

- Gravitational effects are not blocked by intervening matter. For instance, in an eclipse of the moon, the earth is lined up directly between the sun and the moon, but only the sun's light is blocked from reaching the moon, not its gravitational force — if the sun's gravitational force on the moon was blocked in this situation, astronomers would be able to tell because the moon's acceleration would change suddenly. A more subtle but more easily observable example is that the tides are caused by the moon's gravity, and tidal effects can occur on the side of the earth facing away from the moon. Thus, far-off parts of the earth are not prevented from attracting you with their gravity just because there is other stuff between you and them.

- Prospectors sometimes search for underground deposits of dense minerals by measuring the direction of the local gravitational forces, i.e. the direction things fall or the direction a plumb bob hangs. For instance, the gravitational forces in the region to the west of such a deposit would point along a line slightly to the east of the earth's center. Just because the total gravitational force on you points down, that doesn't mean that only the parts of the earth directly below you are attracting you. It's just that the sideways components of all the force vectors acting on you come very close to canceling out.

A cubic centimeter of lava in the earth's mantle, a grain of silica inside Mt. Kilimanjaro, and a flea on a cat in Paris are all attracting you with their gravity. What you feel is the vector sum of all the gravitational forces exerted by all the atoms of our planet, and for that matter by all the atoms in the universe.

When Newton tested his theory of gravity by comparing the orbital acceleration of the moon to the acceleration of a falling apple on earth, he assumed he could compute the earth's force on the apple using the distance from the apple to the earth's center. Was he wrong? After all, it isn't just the earth's center attracting the apple, it's the whole earth. A kilogram of dirt a few feet under his backyard in England would have a much greater force on the apple than a kilogram of molten rock deep under Australia, thousands of miles away. There's really no obvious reason why the force should come out right if you just pretend that the earth's whole mass is concentrated at its center. Also, we know that the earth has some parts that are more dense, and some parts that are less dense. The solid crust, on which we live, is considerably less dense than the molten rock on which it floats. By all rights, the computation of the vector sum of all the forces exerted by all the earth's parts should be a horrendous mess.

Actually, Newton had sound mathematical reasons for treating the earth's mass as if it was concentrated at its center. First, although Newton no doubt suspected the earth's density was nonuniform, he knew that the direction of its total gravitational force was very nearly toward the earth's center. That was strong evidence that the distribution of mass was very symmetric, so that we can think of the earth as being made of many layers, like an onion, with each layer having constant density throughout. (Today there is further evidence for symmetry based on measurements of how the vibrations from earthquakes and nuclear explosions travel through the earth.) Newton then concentrated on the gravitational forces exerted by a single such thin shell, and proved the following mathematical theorem, known as the shell theorem:

> If an object lies outside a thin, uniform shell of mass, then the vector sum of all the gravitational forces exerted by all the parts of the shell is the same as if all the shell's mass was concentrated at its center. If the object lies inside the shell, then all the gravitational forces cancel out exactly.

For terrestrial gravity, each shell acts as though its mass was concentrated at the earth's center, so the final result is the same as if the earth's whole mass was concentrated at its center.

The second part of the shell theorem, about the gravitational forces canceling inside the shell, is a little surprising. Obviously the forces would all cancel out if you were at the exact center of a shell, but why should they still cancel out perfectly if you are inside the shell but off-center? The whole idea might seem academic, since we don't know of any hollow planets in our solar system that astronauts could hope to visit, but actually it's a useful result for understanding gravity within the earth, which is an important issue in geology. It doesn't matter that the earth is not actually hollow. In a mine shaft at a depth of, say, 2 km, we can use the shell theorem to tell us that the outermost 2 km of the earth has no net gravitational effect, and the gravitational force is the same as what would be produced if the remaining, deeper, parts of the earth were all concentrated at its center.

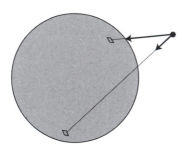

An object outside a spherical shell of mass will feel gravitational forces from every part of the shell — stronger forces from the closer parts and weaker ones from the parts farther away. The shell theorem states that the vector sum of all the forces is the same as if all the mass had been concentrated at the center of the shell.

Discussion Questions

A. If you hold an apple in your hand, does the apple exert a gravitational force on the earth? Is it much weaker than the earth's gravitational force on the apple? Why doesn't the earth seem to accelerate upward when you drop the apple?

B. When astronauts travel from the earth to the moon, how does the gravitational force on them change as they progress?

C. How would the gravity in the first-floor lobby of a massive skyscraper compare with the gravity in an open field outside of the city?

D. In a few billion years, the sun will start undergoing changes that will eventually result in its puffing up into a red giant star. (Near the beginning of this process, the earth's oceans will boil off, and by the end, the sun will probably swallow the earth completely.) As the sun's surface starts to get closer and close to the earth, how will the earth's orbit be affected?

10.5   Weighing the Earth

Let's look more closely at the application of Newton's law of gravity to objects on the earth's surface. Since the earth's gravitational force is the same as if its mass was all concentrated at its center, the force on a falling object of mass m is given by

$$F = G M_{earth} m / r_{earth}^2 \ .$$

The object's acceleration equals F/m, so the object's mass cancels out and we get the same acceleration for all falling objects, as we knew we should:

$$g = G M_{earth} / r_{earth}^2 \ .$$

Newton knew neither the mass of the earth nor a numerical value for the constant G. But if someone could measure G, then it would be possible for the first time in history to determine the mass of the earth! The only way to measure G is to measure the gravitational force between two objects of known mass, but that's an exceedingly difficult task, because the force between any two objects of ordinary size is extremely small. The English physicist Henry Cavendish was the first to succeed, using the apparatus shown in the diagrams. The two larger balls were lead spheres 8 inches in diameter, and each one attracted the small ball near it. The two small balls hung from the ends of a horizontal rod, which itself hung by a thin thread. The frame from which the larger balls hung could be rotated by hand about a vertical axis, so that for instance the large ball on the right would pull its

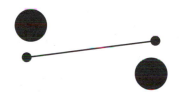

Cavendish's apparatus viewed from the side, and a simplified version viewed from above. The two large balls are fixed in place, but the rod from which the two small balls hang is free to twist under the influence of the gravitational forces.

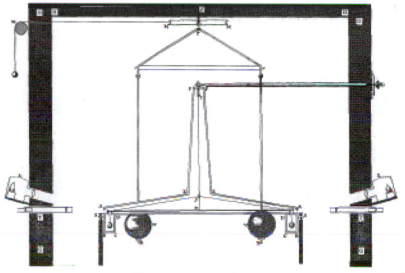

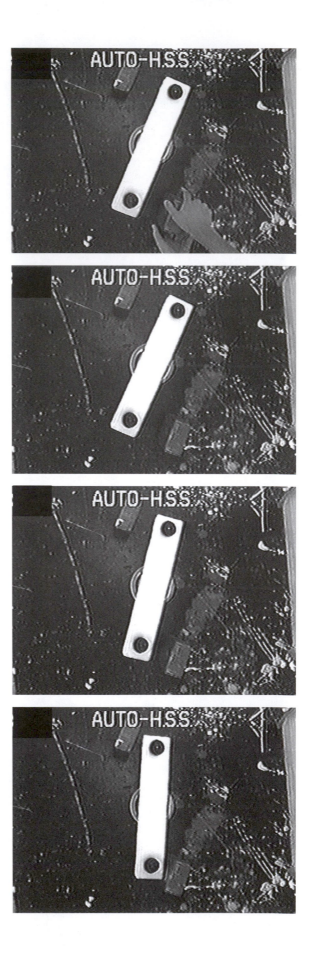

My student Narciso Guzman built this version of the Cavendish experiment in his garage, from a description on the Web at www.fourmilab.to. Two steel balls sit near the ends of a piece of styrofoam, which is suspending from a ladder by fishing line (not visible in this photo). To make vibrations die out more quickly, a small piece of metal from a soda can is attached underneath the styrofoam arm, sticking down into a bowl of water. (The arm is not resting on the bowl.)

The sequence of four video frames on the right shows the apparatus in action. Initially (top), lead bricks are inserted near the steel balls. They attract the balls, and the arm begins to rotate counterclockwise.

The main difficulties in this experiment are isolating the apparatus from vibrations and air currents. Narciso had to leave the room while the camcorder ran. Also, it is helpful if the apparatus can be far from walls or furniture that would create gravitational forces on it.

neighboring small ball toward us and while the small ball on the left would be pulled away from us. The thread from which the small balls hung would thus be twisted through a small angle, and by calibrating the twist of the thread with known forces, the actual gravitational force could be determined. Cavendish set up the whole apparatus in a room of his house, nailing all the doors shut to keep air currents from disturbing the delicate apparatus. The results had to be observed through telescopes stuck through holes drilled in the walls. Cavendish's experiment provided the first numerical values for G and for the mass of the earth. The presently accepted value of G is 6.67×10^{-11} N·m^2/kg^2.

The facing page shows a modern-day Cavendish experiment constructed by one of my students.

Knowing G not only allowed the determination of the earth's mass but also those of the sun and the other planets. For instance, by observing the acceleration of one of Jupiter's moons, we can infer the mass of Jupiter. The following table gives the distances of the planets from the sun and the masses of the sun and planets. (Other data are given in the back of the book.)

	average distance from the sun, in units of the earth's average distance from the sun	mass, in units of the earth's mass
sun	—	330,000
mercury	0.38	.056
venus	.72	.82
earth	1	1
mars	1.5	.11
jupiter	5.2	320
saturn	9.5	95
uranus	19	14
neptune	30	17
pluto	39	.002

Discussion Questions

A. It would have been difficult for Cavendish to start designing an experiment without at least some idea of the order of magnitude of G. How could he estimate it in advance to within a factor of 10?

B. Fill in the details of how one would determine Jupiter's mass by observing the acceleration of one of its moons. Why is it only necessary to know the acceleration of the moon, not the actual force acting on it? Why don't we need to know the mass of the moon? What about a planet that has no moons, such as Venus — how could its mass be found?

C. The gravitational constant G is very difficult to measure accurately, and is the least accurately known of all the fundamental numbers of physics such as the speed of light, the mass of the electron, etc. But that's in the mks system, based on the meter as the unit of length, the kilogram as the unit of mass, and the second as the unit of distance. Astronomers sometimes use a different system of units, in which the unit of distance, called the astronomical unit or a.u., is the radius of the earth's orbit, the unit of mass is the mass of the sun, and the unit of time is the year (i.e. the time required for the earth to orbit the sun). In this system of units, G has a precise numerical value simply as a matter of definition. What is it?

10.6* Evidence for Repulsive Gravity

Until recently, physicists thought they understood gravity fairly well. Einstein had modified Newton's theory, but certain characteristics of gravitational forces were firmly established. For one thing, they were always attractive. If gravity always attracts, then it is logical to ask why the universe doesn't collapse. Newton had answered this question by saying that if the universe was infinite in all directions, then it would have no geometric center toward which it would collapse; the forces on any particular star or planet exerted by distant parts of the universe would tend to cancel out by symmetry. More careful calculations, however, show that Newton's universe would have a tendency to collapse on smaller scales: any part of the universe that happened to be slightly more dense than average would contract further, and this contraction would result in stronger gravitational forces, which would cause even more rapid contraction, and so on.

When Einstein overhauled gravity, the same problem reared its ugly head. Like Newton, Einstein was predisposed to believe in a universe that was static, so he added a special repulsive term to his equations, intended to prevent a collapse. This term was not associated with any attraction of mass for mass, but represented merely an overall tendency for space itself to expand unless restrained by the matter that inhabited it. It turns out that Einstein's solution, like Newton's, is unstable. Furthermore, it was soon discovered observationally that the universe was expanding, and this was interpreted by creating the Big Bang model, in which the universe's current expansion is the aftermath of a fantastically hot explosion. An expanding universe, unlike a static one, was capable of being explained with Einstein's equations, without any repulsion term. The universe's expansion would simply slow down over time due to the attractive gravitational forces. After these developments, Einstein said woefully that adding the repulsive term, known as the cosmological constant, had been the greatest blunder of his life.

This was the state of things until 1999, when evidence began to turn up that the universe's expansion has been speeding up rather than slowing down! The first evidence came from using a telescope as a sort of time

> Book 3, section 3.5 presents some of the evidence for the Big Bang.

machine: light from a distant galaxy may have taken billions of years to reach us, so we are seeing it as it was far in the past. Looking back in time, astronomers saw the universe expanding at speeds that ware lower, rather than higher. At first they were mortified, since this was exactly the opposite of what had been expected. The statistical quality of the data was also not good enough to constute ironclad proof, and there were worries about systematic errors. The case for an accelerating expansion has however been nailed down by high-precision mapping of the dim, sky-wide afterglow of the Big Bang, known as the cosmic microwave background. Some theorists have proposed reviving Einstein's cosmological constant to account for the acceleration, while others believe it is evidence for a mysterious form of matter which exhibits gravitational repulsion. The generic term for this unknown stuff is "dark energy." Some recent ideas on this topic can be found in the January 2001 issue of Scientific American, which is available online at

http://www.sciam.com/issue.cfm?issueDate=Jan-01 .

Note added February 15, 2003

The microwave background measurements referred to above have been improved on by a space probe called WMAP, and there is no longer much room for doubt about the repulsion. An article describing these results, from the front page of the New York times on February 12, 2003, is available online at

http://www.nytimes.com/2003/02/12/science/12COSM.html

(free registration required).

Astronomers consider themselves to have entered a new era of high-precision cosmology. The WMAP probe, for example, has measured the age of the universe to be 13.7 ± 0.2 billion years, a figure that could previously be stated only as a fuzzy range from 10 to 20 billion. We know that only 4% of the universe is atoms, with another 23% consisting of unknown subatomic particles, and 73% of dark energy. It's more than a little ironic to know about so many things with such high precision, and yet to know virtually nothing about their nature. For instance, we know that precisely 96% of the universe is something other than atoms, but we know precisely nothing about what something is.

The WMAP probe's map of the cosmic microwave background is like a "baby picture" of the universe.

Summary

Selected Vocabulary

ellipse	a flattened circle; one of the conic sections
conic section	a curve formed by the intersection of a plane and an infinite cone
hyperbola	another conic section; it does not close back on itself
period..............................	the time required for a planet to complete one orbit; more generally, the time for one repetition of some repeating motion
focus................................	one of two special points inside an ellipse: the ellipse consists of all points such that the sum of the distances to the two foci equals a certain number; a hyperbola also has a focus

Notation

G......................................	the constant of proportionality in Newton's law of gravity; the gravitational force of attraction between two 1-kg spheres at a center-to-center distance of 1 m

Summary

Kepler deduced three empirical laws from data on the motion of the planets:

Kepler's elliptical orbit law: The planets orbit the sun in elliptical orbits with the sun at one focus.

Kepler's equal-area law: The line connecting a planet to the sun sweeps out equal areas in equal amounts of time.

Kepler's law of periods: The time required for a planet to orbit the sun is proportional to the long axis of the ellipse raised to the 3/2 power. The constant of proportionality is the same for all the planets.

Newton was able to find a more fundamental explanation for these laws. *Newton's law of gravity* states that the magnitude of the attractive force between any two objects in the universe is given by

$$F = Gm_1 m_2 / r^2 \ .$$

Weightlessness of objects in orbit around the earth is only apparent. An astronaut inside a spaceship is simply falling along with the spaceship. Since the spaceship is falling out from under the astronaut, it appears as though there was no gravity accelerating the astronaut down toward the deck.

Gravitational forces, like all other forces, add like vectors. A gravitational force such as we ordinarily feel is the vector sum of all the forces exerted by all the parts of the earth. As a consequence of this, Newton proved the *shell theorem* for gravitational forces:

If an object lies outside a thin, uniform shell of mass, then the vector sum of all the gravitational forces exerted by all the parts of the shell is the same as if all the shell's mass was concentrated at its center. If the object lies inside the shell, then all the gravitational forces cancel out exactly.

Homework Problems

1 ✓. Roy has a mass of 60 kg. Laurie has a mass of 65 kg. They are 1.5 m apart.
(a) What is the magnitude of the gravitational force of the earth on Roy?
(b) What is the magnitude of Roy's gravitational force on the earth?
(c) What is the magnitude of the gravitational force between Roy and Laurie?
(d) What is the magnitude of the gravitational force between Laurie and the sun?

2 ✓. During a solar eclipse, the moon, earth and sun all lie on the same line, with the moon between the earth and sun. Define your coordinates so that the earth and moon lie at greater x values than the sun. For each force, give the correct sign as well as the magnitude. (a) What force is exerted on the moon by the sun? (b) On the moon by the earth? (c) On the earth by the sun? (d) What total force is exerted on the sun? (e) On the moon? (f) On the earth?

3 ✓. Suppose that on a certain day there is a crescent moon, and you can tell by the shape of the crescent that the earth, sun and moon form a triangle with a 135° interior angle at the moon's corner. What is the magnitude of the total gravitational force of the earth and the sun on the moon?

earth

sun   moon

4 ✓. How high above the Earth's surface must a rocket be in order to have 1/100 the weight it would have at the surface? Express your answer in units of the radius of the Earth.

5 ✓. The star Lalande 21185 was found in 1996 to have two planets in roughly circular orbits, with periods of 6 and 30 years. What is the ratio of the two planets' orbital radii?

6. In a Star Trek episode, the Enterprise is in a circular orbit around a planet when something happens to the engines. Spock then tells Kirk that the ship will spiral into the planet's surface unless they can fix the engines. Is this scientifically correct? Why?

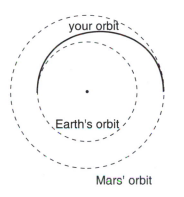

Problem 8.

7. (a✔) Suppose a rotating spherical body such as a planet has a radius r and a uniform density ρ, and the time required for one rotation is T. At the surface of the planet, the apparent acceleration of a falling object is reduced by acceleration of the ground out from under it. Derive an equation for the apparent acceleration of gravity, g, at the equator in terms of r, ρ, T, and G.

(b✔) Applying your equation from (a), by what fraction is your apparent weight reduced at the equator compared to the poles, due to the Earth's rotation?

(c✔) Using your equation from (a), derive an equation giving the value of T for which the apparent acceleration of gravity becomes zero, i.e. objects can spontaneously drift off the surface of the planet. Show that T only depends on ρ, and not on r.

(d) Applying your equation from (c), how long would a day have to be in order to reduce the apparent weight of objects at the equator of the Earth to zero? [Answer: 1.4 hours]

(e) Observational astronomers have recently found objects they called pulsars, which emit bursts of radiation at regular intervals of less than a second. If a pulsar is to be interpreted as a rotating sphere beaming out a natural "searchlight" that sweeps past the earth with each rotation, use your equation from (c) to show that its density would have to be much greater than that of ordinary matter.

(f) Theoretical astronomers predicted decades ago that certain stars that used up their sources of energy could collapse, forming a ball of neutrons with the fantastic density of ~10^{17} kg/m³. If this is what pulsars really are, use your equation from (c) to explain why no pulsar has ever been observed that flashes with a period of less than 1 ms or so.

8✔. You are considering going on a space voyage to Mars, in which your route would be half an ellipse, tangent to the Earth's orbit at one end and tangent to Mars' orbit at the other. Your spacecraft's engines will only be used at the beginning and end, not during the voyage. How long would the outward leg of your trip last? (Assume the orbits of Earth and Mars are circular.)

9.★ (a) If the earth was of uniform density, would your weight be increased or decreased at the bottom of a mine shaft? Explain. (b) In real life, objects weight slightly more at the bottom of a mine shaft. What does that allow us to infer about the Earth?

10 S. Ceres, the largest asteroid in our solar system, is a spherical body with a mass 6000 times less than the earth's, and a radius which is 13 times smaller. If an astronaut who weighs 400 N on earth is visiting the surface of Ceres, what is her weight?

11 S. Prove, based on Newton's laws of motion and Newton's law of gravity, that all falling objects have the same acceleration if they are dropped at the same location on the earth and if other forces such as friction are unimportant. Do not just say, "g=9.8 m/s² -- it's constant." You are supposed to be *proving* that g should be the same number for all objects.

Problem 12.

12 S. The figure shows an image from the Galileo space probe taken during its August 1993 flyby of the asteroid Ida. Astronomers were surprised when Galileo detected a smaller object orbiting Ida. This smaller object, the only known satellite of an asteroid in our solar system, was christened Dactyl, after the mythical creatures who lived on Mount Ida, and who protected the infant Zeus. For scale, Ida is about the size and shape of Orange County, and Dactyl the size of a college campus. Galileo was unfortunately unable to measure the time, T, required for Dactyl to orbit Ida. If it had, astronomers would have been able to make the first accurate determination of the mass and density of an asteroid. Find an equation for the density, ρ, of Ida in terms of Ida's known volume, V, the known radius, r, of Dactyl's orbit, and the lamentably unknown variable T. (This is the same technique that was used successfully for determining the masses and densities of the planets that have moons.)

13 ∫√. If a bullet is shot straight up at a high enough velocity, it will never return to the earth. This is known as the escape velocity. We will discuss escape velocity using the concept of energy in the next book of the series, but it can also be gotten at using straightforward calculus. In this problem, you will analyze the motion of an object of mass m whose initial velocity is *exactly* equal to escape velocity. We assume that it is starting from the surface of a spherically symmetric planet of mass M and radius b. The trick is to guess at the general form of the solution, and then determine the solution in more detail. Assume (as is true) that the solution is of the form $r = kt^p$, where r is the object's distance from the center of the planet at time t, and k and p are constants. (a) Find the acceleration, and use Newton's second law and Newton's law of gravity to determine k and p. You should find that the result is independent of m. (b) What happens to the velocity as t approaches infinity? (c) Determine escape velocity from the Earth's surface.

14. Astronomers have recently observed stars orbiting at very high speeds around an unknown object near the center of our galaxy. For stars orbiting at distances of about 10^{14} m from the object, the orbital velocities are about 10^6 m/s. Assuming the orbits are circular, estimate the mass of the object, in units of the mass of the sun, 2×10^{30} kg. If the object was a tightly packed cluster of normal stars, it should be a very bright source of light. Since no visible light is detected coming from it, it is instead believed to be a supermassive black hole.

15 S. Astronomers have detected a solar system consisting of three planets orbiting the star Upsilon Andromedae. The planets have been named b, c, and d. Planet b's average distance from the star is 0.059 A.U., and planet c's average distance is 0.83 A.U., where an astronomical unit or A.U. is defined as the distance from the Earth to the sun. For technical reasons, it is possible to determine the ratios of the planets' masses, but their masses cannot presently be determined in absolute units. Planet c's mass is 3.0 times that of planet b. Compare the star's average gravitational force on planet c with its average force on planet b. [Based on a problem by Arnold Arons.]

16 S. Some communications satellites are in orbits called geosynchronous: the satellite takes one day to orbit the earth from west to east, so that as the earth spins, the satellite remains above the same point on the equator. What is such a satellite's altitude above the surface of the earth?

17 S. As is discussed in more detail in section 5.1 of book 2, tidal interactions with the earth are causing the moon's orbit to grow gradually larger. Laser beams bounced off of a mirror left on the moon by astronauts have allowed a measurement of the moon's rate of recession, which is about 1 cm per year. This means that the gravitational force acting between earth and moon is decreasing. By what fraction does the force decrease with each 27-day orbit? [Hint: If you try to calculate the two forces and subtract, your calculator will probably give a result of zero due to rounding. Instead, reason about the fractional amount by which the quantity $1/r^2$ will change. As a warm-up, you may wish to observe the percentage change in $1/r^2$ that results from changing r from 1 to 1.01. Based on a problem by Arnold Arons.]

18. Suppose that we inhabited a universe in which, instead of Newton's law of gravity, we had $F = k\sqrt{m_1 m_2}/r^2$, where k is some constant with different units than G. (The force is still attractive.) However, we assume that $a=F/m$ and the rest of Newtonian physics remains true, and we use $a=F/m$ to define our mass scale, so that, e.g., a mass of 2 kg is one which exhibits half the acceleration when the same force is applied to it as to a 1 kg mass. (a) Is this new law of gravity consistent with Newton's third law? (b) Suppose you lived in such a universe, and you dropped two unequal masses side by side. What would happen? (c) Numerically, suppose a 1.0-kg object falls with an acceleration of 10 m/s². What would be the acceleration of a rain drop with a mass of 0.1 g? Would you want to go out in the rain? (d) If a falling object broke into two unequal pieces while it fell, what would happen? (e) Invent a law of gravity that results in behavior that is the opposite of what you found in part b. [Based on a problem by Arnold Arons.]

19 S. (a) A certain vile alien gangster lives on the surface of an asteroid, where his weight is 0.20 N. He decides he needs to lose weight without reducing his consumption of princesses, so he's going to move to a different asteroid where his weight will be 0.10 N. The real estate agent's database has asteroids listed by mass, however, not by surface gravity. Assuming that all asteroids are spherical and have the same density, how should the mass of his new asteroid compare with that of his old one? (b) Jupiter's mass is 318 times the Earth's, and its gravity is about twice Earth's. Is this consistent with the results of part a? If not, how do you explain the discrepancy?

20. Where would an object have to be located so that it would experience zero total gravitational force from the earth and moon?

21✓. The planet Uranus has a mass of 8.68×10^{25} kg and a radius of 2.56×10^4 km. The figure shows the relative sizes of Uranus and Earth. (a) Compute the ratio g_U/g_E, where g_U is the strength of the gravitational field at the surface of Uranus and g_E is the corresponding quantity at the surface of the Earth. (b) What is surprising about this result? How do you explain it?

Problem 21. The sizes of Uranus and Earth are compared. The image of Uranus is from the Voyager 2 probe, and the photo of the earth was taken by the Apollo 11 astronauts.

Exercises

Exercise 0A: Models and Idealization

Equipment:
> coffee filters
> ramps (one per group)
> balls of various sizes
> sticky tape
> vacuum pump and "guinea and feather" apparatus (one)

The motion of falling objects has been recognized since ancient times as an important piece of physics, but the motion is inconveniently fast, so in our everyday experience it can be hard to tell exactly what objects are doing when they fall. In this exercise you will use several techniques to get around this problem and study the motion. Your goal is to construct a scientific *model* of falling. A model means an explanation that makes testable predictions. Often models contain simplifications or idealizations that make them easier to work with, even though they are not strictly realistic.

1. One method of making falling easier to observe is to use objects like feathers that we know from everyday experience will not fall as fast. You will use coffee filters, in stacks of various sizes, to test the following two hypotheses and see which one is true, or whether neither is true:

> Hypothesis 1A: When an object is dropped, it rapidly speeds up to a certain natural falling speed, and then continues to fall at that speed. The falling speed is *proportional* to the object's weight. (A proportionality is not just a statement that if one thing gets bigger, the other does too. It says that if one becomes three times bigger, the other also gets three times bigger, etc.)

> Hypothesis 1B: Different objects fall the same way, regardless of weight.

Test these hypotheses and discuss your results with your instructor.

2. A second way to slow down the action is to let a ball roll down a ramp. The steeper the ramp, the closer to free fall. Based on your experience in part 1, write a hypothesis about what will happen when you race a heavier ball against a lighter ball down the same ramp, starting them both from rest.

Hypothesis:_____

Show your hypothesis to your instructor, and then test it.

You have probably found that falling was more complicated than you thought! Is there more than one factor that affects the motion of a falling object? Can you imagine certain idealized situations that are simpler? Try to agree verbally with your group on an informal model of falling that can make predictions about the experiments described in parts 3 and 4.

3. You have three balls: a standard "comparison ball" of medium weight, a light ball, and a heavy ball. Suppose you stand on a chair and (a) drop the light ball side by side with the comparison ball, then (b) drop the heavy ball side by side with the comparison ball, then (c) join the light and heavy balls together with sticky tape and drop them side by side with the comparison ball.

Use your model to make a prediction:_____

Test your prediction.

4. Your instructor will pump nearly all the air out of a chamber containing a feather and a heavier object, then let them fall side by side in the chamber.

Use your model to make a prediction:_____

Exercise 1A: Scaling Applied to Leaves

Equipment:

leaves of three sizes, having roughly similar proportions of length, width, and thickness
(example: blades of grass, large ficus leaves, and agave leaves)

balance

1. Each group will have one leaf, and should measure its surface area and volume, and determine its surface-to-volume ratio (surface area divided by volume). For consistency, every group should use units of cm^2 and cm^3, and should only find the area of one side of the leaf. The area can be found by tracing the area of the leaf on graph paper and counting squares. The volume can be found by weighing the leaf and assuming that its density is 1 g/cm^3, which is nearly true since leaves are mostly water. Write your results on the board for comparison with the other groups' numbers.

2. Both the surface area and the volume are bigger for bigger leaves, but what about the surface to volume ratios? What implications would this have for the plants' abilities to survive in different environments?

Exercise 2A: Changing Velocity

This exercise involves Michael Johnson's world-record 200-meter sprint in the 1996 Olympics. The table gives the distance he has covered at various times. (The data are made up, except for his total time.) Each group is to find a value of $\Delta x/\Delta t$ between two specified instants, with the members of the group checking each other's answers. We will then compare everyone's results and discuss how this relates to velocity.

	t (s)	x (m)
A	10.200	100.0000
B	10.210	100.0990
C	10.300	100.9912
D	11.200	110.0168
E	19.320	200.0000

group 1:   Find $\Delta x/\Delta t$ using points A and B.
group 2:   Find $\Delta x/\Delta t$ using points A and C.
group 3:   Find $\Delta x/\Delta t$ using points A and D.
group 4:   Find $\Delta x/\Delta t$ using points A and E.

Exercise 3A: Reasoning with Ratios and Powers

Equipment:

      ping-pong balls and paddles

      two-meter sticks

You have probably bounced a ping pong ball straight up and down in the air. The time between hits is related to the height to which you hit the ball. If you take twice as much time between hits, how many times higher do you think you will have to hit the ball? Write down your hypothesis:_____

Your instructor will first beat out a tempo of 240 beats per minute (four beats per second), which you should try to match with the ping-pong ball. Measure the height to which the ball rises:_____

Now try it at 120 beats per minute:_____

Compare your hypothesis and your results with the rest of the class.

Exercise 4A: Force and Motion

Equipment:
> 2-meter pieces of butcher paper
> wood blocks with hooks
> string
> masses to put on top of the blocks to increase friction
> spring scales (preferably calibrated in Newtons)

Suppose a person pushes a crate, sliding it across the floor at a certain speed, and then repeats the same thing but at a higher speed. This is essentially the situation you will act out in this exercise. What do you think is different about her force on the crate in the two situations? Discuss this with your group and write down your hypothesis:

_____

1. First you will measure the amount of friction between the wood block and the butcher paper when the wood and paper surfaces are slipping over each other. The idea is to attach a spring scale to the block and then slide the butcher paper under the block while using the scale to keep the block from moving with it. Depending on the amount of force your spring scale was designed to measure, you may need to put an extra mass on top of the block in order to increase the amount of friction. It is a good idea to use long piece of string to attach the block to the spring scale, since otherwise one tends to pull at an angle instead of directly horizontally.

First measure the amount of friction force when sliding the butcher paper as slowly as possible:_____

Now measure the amount of friction force at a significantly higher speed, say 1 meter per second. (If you try to go too fast, the motion is jerky, and it is impossible to get an accurate reading.)
_____

Discuss your results. Why are we justified in assuming that the string's force on the block (i.e. the scale reading) is the same amount as the paper's frictional force on the block?

2. Now try the same thing but with the block moving and the paper standing still. Try two different speeds.

Do your results agree with your original hypothesis? If not, discuss what's going on. How does the block "know" how fast to go?

Exercise 4B: Interactions

Equipment:
- neodymium disc magnets (3/group)
- compass
- triple-arm balance (2/group)
- clamp and 50-cm vertical rod for holding balance up
- string
- tape
- scissors

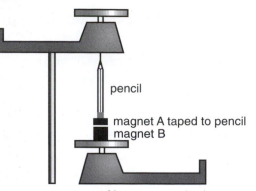

pencil

magnet A taped to pencil
magnet B

Your goal in this exercise is to compare the forces two magnets exert on each other, i.e. to compare magnet A's force on magnet B to magnet B's force on magnet A. Magnet B will be made out of two of the small disc magnets put together, so it is twice as strong as magnet A.

1. Note that these magnets are extremely strong! Being careful not to pinch your skin, put two disc magnets together to make magnet B.

2. Familiarize yourself with how the magnets behave. In addition to magnets A and B, there are two other magnets that can come into play. The compass needle itself is a magnet, and the planet earth is a magnet. Ordinarily the compass needle twists around under the influence of the earth, but the disc magnets are very strong close up, so if you bring them within a few cm of the compass, the compass is essentially just responding to them. Investigate how different parts of magnets A and B interact with the compass, and label them appropriately. Investigate how magnets A and B can attract or repel one another.

3. You are ready to form a hypothesis about the following situation. Suppose we set up two balances as shown in the figure. The magnets are not touching. The top magnet is hanging from a hook underneath the pan, giving the same result as if it was on top of the pan. Make sure it is hanging under the *center* of the pan. You will want to make sure the magnets are pulling on each other, not pushing each other away, so that the top magnet will stay in one place.

The balances will not show the magnets' true weights, because the magnets are exerting forces on each other. The top balance will read a higher number than it would without any magnetic forces, and the bottom balance will have a lower than normal reading. The difference between each magnet's true weight and the reading on the bal-

ance gives a measure of how strongly the magnet is being pushed or pulled by the other magnet.

How do you think the amount of pushing or pulling experienced by the two magnets will compare? In other words, which reading will change more, or will they change by the same amount? Write down a hypothesis:_____

Before going on to part 4, discuss your hypothesis with your instructor.

4. Now set up the experiment described above with two balances. Since we are interested in the changse in the scale readings caused by the magnetic forces, you will need to take a total of four scale readings: one pair with the balances separated and one pair with the magnets close together as shown in the figure above.

When the balances are together and the magnetic forces are acting, it is not possible to get both balances to reach equilibrium at the same time, because sliding the weights on one balance can cause its magnet to move up or down, tipping the other balance. Therefore, while you take a reading from one balance, you need to immobilize the other in the horizontal position by taping its tip so it points exactly at the zero mark.

You will also probably find that as you slide the weights, the pointer swings suddenly to the opposite side, but you can never get it to be stable in the middle (zero) position. Try bringing the pointer manually to the zero position and then releasing it. If it swings up, you're too low, and if it swings down, you're too high. Search for the dividing line between the too-low region and the too-high region.

If the changes in the scale readings are very small (say a few grams or less), you need to get the magnets closer together. It should be possible to get the scale readings to change by large amounts (up to 10 or 20 g).

Exercise 5A: Friction

Equipment:

     2-meter pieces of butcher paper
     wood blocks with hooks
     string
     masses to put on top of the blocks to increase friction
     spring scales (preferably calibrated in Newtons)

1. Using the same equipment as in exercise 4A, test the statement that kinetic friction is approximately independent of velocity.

2. Test the statement that kinetic friction is independent of surface area.

Exercise 8A: Vectors and Motion

Each diagram on the right shows the motion of an object. Each dot is one location of the object at one moment in time. The time interval from one dot to the next is always the same.

1. Suppose the object in diagram 1 is moving from the top left to the bottom right. Figure out what kind of force is acting on it. Does the force always have the same magnitude? The same direction?

Invent a physical situation that this diagram could represent.

What if you reinterpret the diagram, and reverse the object's direction of motion?

2. What kind of force is acting in diagram 2?

Invent a physical situation that diagram 2 could represent.

3. What kind of force is acting in diagram 3?

Invent a physical situation.

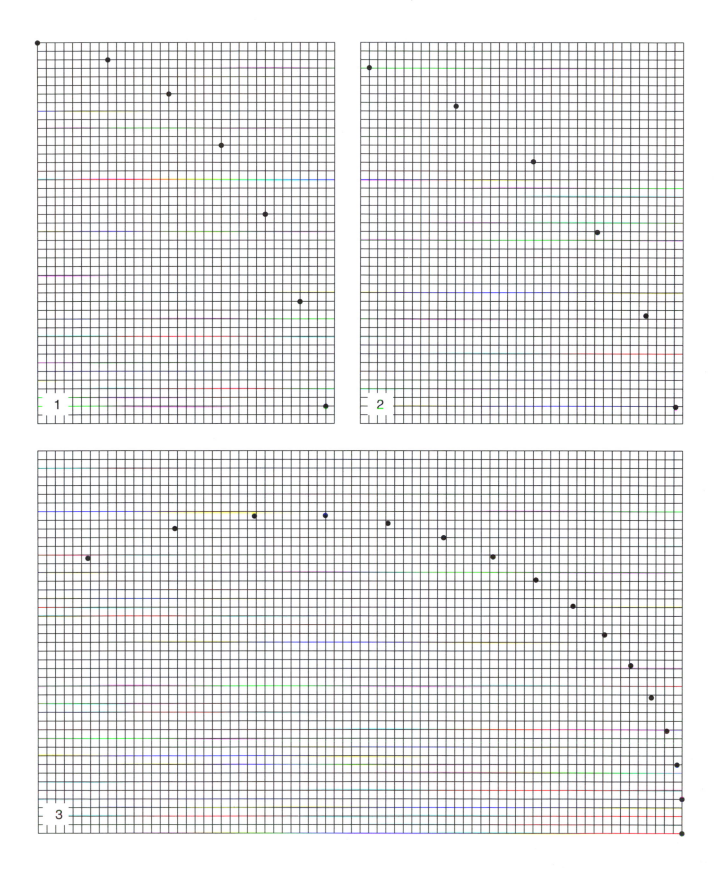

Exercise 10A: The Shell Theorem

This exercise is an approximate numerical test of the shell theorem. There are seven masses A-G, each being one kilogram. Masses A-E, each one meter from the center, form a shape like two Egyptian pyramids joined at their bases; this is a rough approximation to a six-kilogram spherical shell of mass. Mass G is five meters from the center of the main group. The class will divide into six groups and split up the work required in order to calculate the vector sum of the six gravitational forces exerted on mass G. Depending on the size of the class, more than one group may be assigned to deal with the contribution of the same mass to the total force, and the redundant groups can check each other's results.

1. Discuss as a class what can be done to simplify the task of calculating the vector sum, and how to organize things so that each group can work in parallel with the others.

2. Each group should write its results on the board in units of piconewtons, retaining six significant figures of precision.

3. The class will determine the vector sum and compare with the result that would be obtained with the shell theorem.

Solutions to Selected Problems

Chapter 0

6. $134 \text{ mg} \times \dfrac{10^{-3} \text{ g}}{1 \text{ mg}} \times \dfrac{10^{-3} \text{ kg}}{1 \text{ g}} = 1.34 \times 10^{-4} \text{ kg}$

8. (a) Let's do 10.0 g and 1000 g. The arithmetic mean is 505 grams. It comes out to be 0.505 kg, which is consistent. (b) The geometric mean comes out to be 100 g or 0.1 kg, which is consistent. (c) If we multiply meters by meters, we get square meters. Multiplying grams by grams should give square grams! This sounds strange, but it makes sense. Taking the square root of square grams (g^2) gives grams again. (d) No. The superduper mean of two quantities with units of grams wouldn't even be something with units of grams! Related to this shortcoming is the fact that the superduper mean would fail the kind of consistency test carried out in the first two parts of the problem.

Chapter 1

10. $1 \text{ mm}^2 \times \left(\dfrac{1 \text{ cm}}{10 \text{ mm}}\right)^2 = 10^{-2} \text{ cm}^2$

11. The bigger scope has a diameter that's ten times greater. Area scales as the square of the linear dimensions, so its light-gathering power is a hundred times greater (10x10).

12. Since they differ by two steps on the Richter scale, the energy of the bigger quake is 10000 times greater. The wave forms a hemisphere, and the surface area of the hemisphere over which the energy is spread is proportional to the square of its radius. If the amount of vibration was the same, then the surface areas much be in the ratio of 10000:1, which means that the ratio of the radii is 100:1.

Chapter 2

4. $1 \text{ light-year} = v\Delta t$

$= \left(3.0\text{x}10^8 \text{ m/s}\right)(1 \text{ year})$

$\times \left(\dfrac{365 \text{ days}}{1 \text{ year}}\right)\left(\dfrac{24 \text{ hours}}{1 \text{ day}}\right)\left(\dfrac{3600 \text{ s}}{1 \text{ hour}}\right)$

$= 9.5\text{x}10^{15} \text{ m}$

5. Velocity is relative, so having to lean tells you nothing about the train's velocity. Fullerton is moving at a huge speed relative to Beijing, but that doesn't produce any noticeable effect in either city. The fact that you have to lean tells you that the train is accelerating.

7. To the person riding the moving bike, bug A is simply going in circles. The only difference between the motions of the two wheels is that one is traveling through space, but motion is relative, so this doesn't have any effect on the bugs. It's equally hard for each of them.

10. In one second, the ship moves v meters to the east, and the person moves v meters north relative to the deck. Relative to the water, he traces the diagonal of a triangle whose length is given by the Pythagorean theorem, $(v^2 + v^2)^{1/2} = 2^{1/2}v$. Relative to the water, he is moving at a 45-degree angle between north and east.

Chapter 3

14.

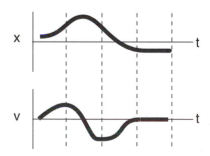

15. Taking g to be 10 m/s, the bullet loses 10 m/s of speed every second, so it will take 10 s to come to a stop, and then another 10 s to come back down, for a total of 20 s.

16. $\Delta x = \frac{1}{2}at^2$, so for a fixed value of Δx, we have $t \propto 1/\sqrt{a}$. Decreasing a by a factor of 3 means that t will increase by a factor of $\sqrt{3}$.

17. $v = \dfrac{dx}{dt}$

$\quad = 10 - 3t^2$

$\quad a = \dfrac{dv}{dt}$

$\quad\quad = -6t$

$\quad\quad = -18 \text{ m/s}^2$

18. (a) Solving $\Delta x = \frac{1}{2}at^2$ for a, we find $a=2\Delta x/t^2=5.51$ m/s². (b) $v=\sqrt{2a\Delta x}=66.6$ m/s. (c) The actual car's final velocity is less than that of the idealized constant-acceleration car. If the real car and the idealized car covered the quarter mile in the same time but the real car was moving more slowly at the end than the idealized one, the real car must have been going faster than the idealized car at the beginning of the race. The real car apparently has a greater acceleration at the beginning, and less acceleration at the end. This make sense, because every car has some maximum speed, which is the speed beyond which it cannot accelerate.

19. Since the lines are at intervals of one m/s and one second, each box represents one meter. From $t=0$ to $t=2$ s, the area under the curve represents a positive Δx of 6 m. (The triangle has half the area of the 2x6 rectangle it fits inside.) After $t=2$ s, the area above the curve represents negative Δx. To get –6 m worth of area, we need to go out to $t=6$ s, at which point the triangle under the axis has a width of 4 s and a height of 3 m/s, for an area of 6 m (half of 3x4).

20. (a) We choose a coordinate system with positive pointing to the right. Some people might expect that the ball would slow down once it was on the more gentle ramp. This may be true if there is significant friction, but Galileo's experiments with inclined planes showed that when friction is negligible, a ball rolling on a ramp has constant acceleration, not constant speed. The speed stops increasing as quickly once the ball is on the more gentle slope, but it still keeps on increasing. The a-t graph can be drawn by inspecting the slope of the v-t graph.

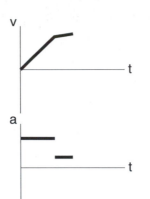

(b) The ball will roll back down, so the second half of the motion is the same as in part a. In the first (rising) half of the motion, the velocity is negative, since the motion is in the opposite direction compared to the positive x axis. The acceleration is again found by inspecting the slope of the v-t graph.

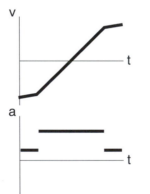

21. This is a case where it's probably easiest to draw the acceleration graph first. While the ball is in the air (bc, de, etc.), the only force acting on it is gravity, so it must have the same, constant acceleration during each hop. Choosing a coordinate system where the positive x axis points up, this becomes a negative acceleration (force in the opposite direction compared to the axis). During the short times between hops when the ball is in contact with the ground (cd, ef, etc.), it experiences a large acceleration, which turns around its velocity very rapidly. These short positive accelerations probably aren't constant, but it's hard to know how they'd really look. We just idealize them as constant accelerations. Similarly, the hand's force on the ball during the time ab is probably not constant, but we can draw it that way, since we don't know how to draw it more realistically. Since our acceleration graph consists of constant-acceleration

segments, the velocity graph must consist of line segments, and the position graph must consist of parabolas.

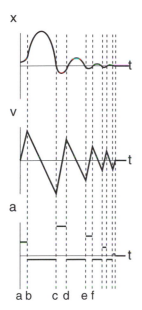

22. We have $v_f^2=2a\Delta x$, so the distance is proportional to the square of the velocity. To get up to half the speed, the ball needs 1/4 the distance, i.e. $L/4$.

Chapter 4

7. $a=\frac{\Delta v}{\Delta t}$, and also $a=\frac{F}{m}$, so

$$\Delta t = \frac{\Delta v}{a}$$

$$= \frac{m\Delta v}{F}$$

$$= \frac{(1000 \text{ kg})(50 \text{ m/s} - 20 \text{ m/s})}{3000 \text{ N}}$$

$$= 10 \text{ s}$$

Chapter 5

14. (a)

top spring's rightward force on connector
...connector's leftward force on top spring

bottom spring's rightward force on connector
...connector's leftward force on bottom spring

hand's leftward force on connector

...connector's rightward force on hand

Looking at the three forces on the connector, we see that the hand's force must be double the force of either spring. The value of $x-x_o$ is the same for both springs and for the arrangement as a whole, so the spring constant must be $2k$. This corresponds to a stiffer spring (more force to produce the same extension).

(b) Forces in which the left spring participates:

hand's leftward force on left spring
...left spring's rightward force on hand

right spring's rightward force on left spring
...left spring's leftward force on right spring

Forces in which the right spring participates:

left spring's leftward force on right spring
...right spring's rightward force on left spring

wall's rightward force on right spring
...right spring's leftward force on wall

Since the left spring isn't accelerating, the total force on it must be zero, so the two forces acting on it must be equal in magnitude. The same applies to the two forces acting on the right spring. The forces between the two springs are connected by Newton's third law, so all eight of these forces must be equal in magnitude. Since the value of $x-x_o$ for the whole setup is double what it is for either spring individually, the spring constant of the whole setup must be $k/2$, which corresponds to a less stiff spring.

16. (a) Spring constants in parallel add, so the spring constant has to be proportional to the cross-sectional area. Two springs in series give half the spring constant, three springs in series give 1/3, and so on, so the spring constant has to be inversely proportional to the length. Summarizing, we have $k \propto A/L$. (b) With the Young's modulus, we have $k=(A/L)E$. The spring constant has units of N/m, so the units of E would have to be N/m^2.

18. (a) The swimmer's acceleration is caused by the water's force on the swimmer, and the swimmer makes a backward force on the water, which accelerates the water backward. (b) The club's normal force on the ball accelerates the ball, and the ball makes a backward normal force on the club, which decelerates the club. (c) The bowstring's normal force accelerates the arrow, and the arrow also makes a backward normal force on the string. This force on the string causes the string to accelerate less rapidly than it would if the bow's force was the only one acting on it. (d) The tracks' backward frictional force slows the locomotive down. The locomotive's forward frictional force causes the whole planet earth to accelerate by a tiny amount, which is too small to measure because the earth's mass is so great.

20. The person's normal force on the box is paired with the box's normal force on the person. The dirt's frictional force on the box pairs with the box's frictional force on the dirt. The earth's gravitational force o n the box matches the box's gravitational force on the earth.

Chapter 6

5. (a) The easiest strategy is to find the time spent aloft, and then find the range. The vertical motion and the horizontal motion are independent. The vertical motion has acceleration $-g$, and the cannonball spends enough time in the air to reverse its vertical velocity component completely, so we have

$$\Delta v_y = v_{yf} - v_{yi}$$
$$= -2v\sin\theta \quad.$$

The time spent aloft is therefore

$$\Delta t = \Delta v_y / a_y$$
$$= 2v\sin\theta / g \quad.$$

During this time, the horizontal distance traveled is

$$R = v_x\Delta t$$
$$= 2 v^2 \sin\theta \cos\theta / g \quad.$$

(b) The range becomes zero at both $\theta=0$ and at $\theta=90°$. The $\theta=0$ case gives zero range because the ball hits the ground as soon as it leaves the mouth of the cannon. A 90 degree angle gives zero range because the cannonball has no horizontal motion.

Chapter 8

8. We want to find out about the velocity vector $\mathbf{v}_{BG}$ of the bullet relative to the ground, so we need to add Annie's velocity relative to the ground $\mathbf{v}_{AG}$ to the bullet's velocity vector $\mathbf{v}_{BA}$ relative to her. Letting the positive x axis be east and y north, we have

$$\mathbf{v}_{BA,x} = (140 \text{ mi/hr}) \cos 45°$$
$$= 100 \text{ mi/hr}$$
$$\mathbf{v}_{BA,y} = (140 \text{ mi/hr}) \sin 45°$$
$$= 100 \text{ mi/hr}$$

and

$$\mathbf{v}_{AG,x} = 0$$
$$\mathbf{v}_{AG,y} = 30 \text{ mi/hr} \quad.$$

The bullet's velocity relative to the ground therefore has components

$$\mathbf{v}_{BG,x} = 100 \text{ mi/hr} \quad \text{and}$$
$$\mathbf{v}_{BG,y} = 130 \text{ mi/hr} \quad.$$

Its speed on impact with the animal is the magnitude of this vector

$$|\mathbf{v}_{BG}| = \sqrt{(100 \text{ mi/hr})^2 + (130 \text{ mi/hr})^2}$$
$$= 160 \text{ mi/hr}$$

(rounded off to 2 significant figures).

9. Since its velocity vector is constant, it has zero acceleration, and the sum of the force vectors acting on it must be zero. There are three forces acting on the plane: thrust, lift, and gravity. We are given the first two, and if we can find the third we can infer its mass. The sum of the y components of the forces is zero, so

$$0 = F_{thrust,y} + F_{lift,y} + F_{W,y}$$
$$= |\mathbf{F}_{thrust}| \sin\theta + |\mathbf{F}_{lift}| \cos\theta - mg \quad.$$

The mass is

$$m = (|\mathbf{F}_{thrust}| \sin\theta + |\mathbf{F}_{lift}| \cos\theta) / g$$
$$= 6.9 \times 10^4 \text{ kg}$$

10. (a) Since the wagon has no acceleration, the total forces in both the x and y directions must be zero. There are three forces acting on the wagon: F_T, F_W, and the normal force from the ground, F_N. If we pick a coordinate system with x being horizontal and y vertical, then the angles of these forces measured counterclockwise from the x axis are 90°-φ, 270°, and 90°+θ, respectively. We have

$$F_{x,total} = F_T\cos(90°-\varphi) + F_W\cos(270°) + F_N\cos(90°+\theta)$$
$$F_{y,total} = F_T\sin(90°-\varphi) + F_W\sin(270°) + F_N\sin(90°+\theta) \quad,$$

which simplifies to

$$0 = F_T \sin\varphi - F_N \sin\theta$$
$$0 = F_T \cos\varphi - F_W + F_N \cos\theta \quad.$$

The normal force is a quantity that we are not given and do not with to find, so we should choose it to eliminate. Solving the first equation for $F_N=(\sin\varphi/\sin\theta)F_T$, we eliminate F_N from the second equation,

$$0 = F_T \cos\varphi - F_W + F_T \sin\varphi \cos\theta/\sin\theta$$

and solve for F_T, finding

$$F_T = \frac{F_W}{\cos\varphi + \sin\varphi \cos\theta / \sin\theta} \quad.$$

Multiplying both the top and the bottom of the fraction by $\sin\theta$, and using the trig identity for $\sin(\theta+\varphi)$ gives the desired result,

$$F_T = \frac{\sin\theta}{\sin(\theta + \varphi)} F_W$$

(b) The case of φ=0, i.e. pulling straight up on the

wagon, results in $F_T=F_W$: we simply support the wagon and it glides up the slope like a chair-lift on a ski slope. In the case of $\varphi=180°-\theta$, F_T becomes infinite. Physically this is because we are pulling directly into the ground, so no amount of force will suffice.

11. (a) If there was no friction, the angle of repose would be zero, so the coefficient of static friction, μ_s, will definitely matter. We also make up symbols θ, m and g for the angle of the slope, the mass of the object, and the acceleration of gravity. The forces form a triangle just like the one in section 8.3, but instead of a force applied by an external object, we have static friction, which is less than $\mu_s F_N$. As in that example, $F_s=mg\sin\theta$, and $F_s<\mu_s F_N$, so

$$mg\sin\theta<\mu_s F_N .$$

From the same triangle, we have $F_N=mg\cos\theta$, so

$$mg\sin\theta < \mu_s mg\cos\theta .$$

Rearranging,

$$\theta < \tan^{-1}\mu_s .$$

(b) Both m and g canceled out, so the angle of repose would be the same on an asteroid.

Chapter 9

5. Each cyclist has a radial acceleration of $v^2/r=5$ m/s^2. The tangential accelerations of cyclists A and B are 375 N/75 kg=5 m/s^2.

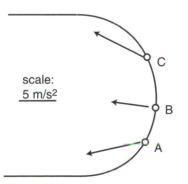

scale:
5 m/s^2

6. (a) The inward normal force must be sufficient to produce circular motion, so

$$F_N = mv^2/r .$$

We are searching for the minimum speed, which is the speed at which the static friction force is just barely able to cancel out the downward gravitational force. The maximum force of static friction is

$$|F_s| = \mu_s F_N ,$$

and this cancels the gravitational force, so

$$|F_s| = mg .$$

Solving these three equations for v gives

$$v = \sqrt{\frac{gr}{\mu_s}} .$$

(b) Greater by a factor of $\sqrt{3}$.

7. The inward force must be supplied by the inward component of the normal force,

$$F_N \sin\theta = mv^2/r .$$

The upward component of the normal force must cancel the downward force of gravity,

$$F_N \cos\theta = mg .$$

Eliminating F_N and solving for θ, we find

$$\theta = \tan^{-1}\left(\frac{v^2}{gr}\right) .$$

Chapter 10

10. Newton's law of gravity tells us that her weight will be 6000 times smaller because of the asteroid's smaller mass, but $13^2=169$ times greater because of its smaller radius. Putting these two factors together gives a reduction in weight by a factor of 6000/169, so her weight will be (400 N)(169)/(6000)=11 N.

11. Newton's law of gravity says $F=Gm_1 m_2/r^2$, and Newton's second law says $F=m_2 a$, so $Gm_1 m_2/r_2=m_2 a$. Since m_2 cancels, a is independent of m_2.

12. Newton's second law gives

$$F= m_D a_D ,$$

where F is Ida's force on Dactyl. Using Newton's universal law of gravity, $F= Gm_1 m_D/r^2$, and the equation $a= v^2/r$ for circular motion, we find

$$Gm_1 m_D/ r^2 = m_D v^2/ r .$$

Dactyl's mass cancels out, giving

$$Gm_1/ r^2 = v^2/ r .$$

Dactyl's velocity equals the circumference of its orbit divided by the time for one orbit: $v=2\pi r/T$. Inserting this in the above equation and solving for m_1, we find

$$m_1 = \frac{4\pi^2 r^3}{GT^2} ,$$

so Ida's density is

$$\rho = m_1/V$$

$$= \frac{4\pi^2 r^3}{GVT^2} .$$

15. Newton's law of gravity depends on the inverse square of the distance, so if the two planets' masses had been equal, then the factor of 0.83/0.059=14 in distance would have caused the force on planet c to be 14^2=2.0x10^2 times weaker. However, planet c's mass is 3.0 times greater, so the force on it is only smaller by a factor of 2.0x10^2/3.0=65.

16. The reasoning is reminiscent of section 10.2. From Newton's second law we have $F=ma=mv^2/r=m(2\pi r/T)^2/r = 4\pi^2 mr/T^2$, and Newton's law of gravity gives $F=GMm/r^2$, where M is the mass of the earth. Setting these expressions equal to each other, we have

$$4\pi^2 mr/T^2 = GMm/r^2 \quad ,$$

which gives

$$r = \sqrt[3]{\frac{GMT^2}{4\pi^2}}$$

$$= 4.22\text{x}10^4 \text{ km} \quad .$$

This is the distance from the center of the earth, so to find the altitude, we need to subtract the radius of the earth. The altitude is 3.58x10^4 km.

17. Any fractional change in r results in double that amount of fractional change in $1/r^2$. For example, raising r by 1% causes $1/r^2$ to go down by very nearly 2%. The fractional change in $1/r^2$ is actually

$$2\times \frac{(1/27)\text{ cm}}{3.84\times10^5 \text{ km}} \times \frac{1 \text{ km}}{10^5 \text{ cm}} = 2\times 10^{-12}$$

19. (a) The asteroid's mass depends on the cube of its radius, and for a given mass the surface gravity depends on r^{-2}. The result is that surface gravity is directly proportional to radius. Half the gravity means half the radius, or one eighth the mass. (b) To agree with a, Earth's mass would have to be 1/8 Jupiter's. We assumed spherical shapes and equal density. Both planets are at least roughly spherical, so the only way out of the contradiction is if Jupiter's density is significantly less than Earth's.

Glossary

Acceleration. The rate of change of velocity; the slope of the tangent line on a *v-t* graph.

Attractive. Describes a force that tends to pull the two participating objects together. Cf. repulsive, oblique.

Center of mass. The balance point of an object.

Component. The part of a velocity, acceleration, or force that is along one particular coordinate axis.

Displacement. (avoided in this book) A name for the symbol Δx .

Fluid. A gas or a liquid.

Fluid friction. A friction force in which at least one of the object is is a fluid (i.e. either a gas or a liquid).

Gravity. A general term for the phenomenon of attraction between things having mass. The attraction between our planet and a human-sized object causes the object to fall.

Inertial frame. A frame of reference that is not accelerating, one in which Newton's first law is true

Kinetic friction. A friction force between surfaces that are slipping past each other.

Light. Anything that can travel from one place to another through empty space and can influence matter, but is not affected by gravity.

Magnitude. The "amount" associated with a vector; the vector stripped of any information about its direction.

Mass. A numerical measure of how difficult it is to change an object's motion.

Matter. Anything that is affected by gravity.

Mks system. The use of metric units based on the meter, kilogram, and second. Example: meters per second is the mks unit of speed, not cm/s or km/hr.

Noninertial frame. An accelerating frame of reference, in which Newton's first law is violated

Nonuniform circular motion. Circular motion in which the magnitude of the velocity vector changes

Normal force. The force that keeps two objects from occupying the same space.

Oblique. Describes a force that acts at some other angle, one that is not a direct repulsion or attraction. Cf. attractive, repulsive.

Operational definition. A definition that states what operations should be carried out to measure the thing being defined.

Parabola. The mathematical curve whose graph has y proportional to x^2.

Radial. Parallel to the radius of a circle; the in-out direction. Cf. tangential.

Repulsive. Describes a force that tends to push the two participating objects apart. Cf. attractive, oblique.

Scalar. A quantity that has no direction in space, only an amount. Cf. vector.

Significant figures. Digits that contribute to the accuracy of a measurement.

Speed. (avoided in this book) The absolute value of or, in more then one dimension, the magnitude of the velocity, i.e. the velocity stripped of any information about its direction

Spring constant. The constant of proportionality between force and elongation of a spring or other object under strain.

Static friction. A friction force between surfaces that are not slipping past each other.

Système International.. Fancy name for the metric system.

Tangential. Tangent to a curve. In circular motion, used to mean tangent to the circle, perpendicular to the radial direction Cf. radial.

Uniform circular motion. Circular motion in which the magnitude of the velocity vector remains constant

Vector. A quantity that has both an amount (magnitude) and a direction in space. Cf. scalar.

Velocity. The rate of change of position; the slope of the tangent line on an *x-t* graph.

Weight. The force of gravity on an object, equal to *mg*.

Mathematical Review

Algebra

Quadratic equation:

The solutions of $ax^2 + bx + c = 0$

are $x = \dfrac{-b \pm \sqrt{b^2 - 4ac}}{2a}$.

Logarithms and exponentials:

$\ln(ab) = \ln a + \ln b$

$e^{a+b} = e^a e^b$

$\ln e^x = e^{\ln x} = x$

$\ln\left(a^b\right) = b \ln a$

Geometry, area, and volume

area of a triangle of base b and height h	$= \frac{1}{2}bh$
circumference of a circle of radius r	$= 2\pi r$
area of a circle of radius r	$= \pi r^2$
surface area of a sphere of radius r	$= 4\pi r^2$
volume of a sphere of radius r	$= \frac{4}{3}\pi r^3$

Trigonometry with a right triangle

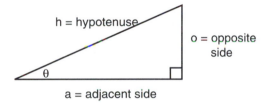

h = hypotenuse
o = opposite side
a = adjacent side
θ

Definitions of the sine, cosine, and tangent:

$\sin\theta = \dfrac{o}{h}$

$\cos\theta = \dfrac{a}{h}$

$\tan\theta = \dfrac{o}{a}$

Pythagorean theorem: $h^2 = a^2 + o^2$

Trigonometry with any triangle

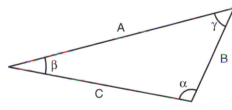

A
γ
β
B
α
C

Law of Sines:

$\dfrac{\sin\alpha}{A} = \dfrac{\sin\beta}{B} = \dfrac{\sin\gamma}{C}$

Law of Cosines:

$C^2 = A^2 + B^2 - 2AB\cos\gamma$

Properties of the derivative and integral (for students in calculus-based courses)

Let f and g be functions of x, and let c be a constant.

Linearity of the derivative:

$\dfrac{d}{dx}(cf) = c\dfrac{df}{dx}$

$\dfrac{d}{dx}(f+g) = \dfrac{df}{dx} + \dfrac{dg}{dx}$

The chain rule:

$\dfrac{d}{dx}f(g(x)) = f'(g(x))g'(x)$

Derivatives of products and quotients:

$\dfrac{d}{dx}(fg) = \dfrac{df}{dx}g + \dfrac{dg}{dx}f$

$\dfrac{d}{dx}\left(\dfrac{f}{g}\right) = \dfrac{f'}{g} - \dfrac{fg'}{g^2}$

Some derivatives:

$\dfrac{d}{dx}x^m = mx^{m-1}$ (except for $m{=}0$)

$\dfrac{d}{dx}\sin x = \cos x$

$\dfrac{d}{dx}\cos x = -\sin x$

$\dfrac{d}{dx}e^x = e^x$

$\dfrac{d}{dx}\ln x = \dfrac{1}{x}$

The fundamental theorem of calculus:

$\displaystyle\int \dfrac{df}{dx}dx = f$

Linearity of the integral:

$\displaystyle\int cf(x)dx = c\int f(x)dx$

$\displaystyle\int \left[f(x) + g(x)\right]dx = \int f(x)dx + \int g(x)dx$

Integration by parts:

$\displaystyle\int f\,dg = fg - \int g\,df$

Trig Tables

θ	sin θ	cos θ	tan θ	θ	sin θ	cos θ	tan θ	θ	sin θ	cos θ	tan θ
0°	0.000	1.000	0.000	30°	0.500	0.866	0.577	60°	0.866	0.500	1.732
1	0.017	1.000	0.017	31	0.515	0.857	0.601	61	0.875	0.485	1.804
2	0.035	0.999	0.035	32	0.530	0.848	0.625	62	0.883	0.469	1.881
3	0.052	0.999	0.052	33	0.545	0.839	0.649	63	0.891	0.454	1.963
4	0.070	0.998	0.070	34	0.559	0.829	0.675	64	0.899	0.438	2.050
5	0.087	0.996	0.087	35	0.574	0.819	0.700	65	0.906	0.423	2.145
6	0.105	0.995	0.105	36	0.588	0.809	0.727	66	0.914	0.407	2.246
7	0.122	0.993	0.123	37	0.602	0.799	0.754	67	0.921	0.391	2.356
8	0.139	0.990	0.141	38	0.616	0.788	0.781	68	0.927	0.375	2.475
9	0.156	0.988	0.158	39	0.629	0.777	0.810	69	0.934	0.358	2.605
10	0.174	0.985	0.176	40	0.643	0.766	0.839	70	0.940	0.342	2.747
11	0.191	0.982	0.194	41	0.656	0.755	0.869	71	0.946	0.326	2.904
12	0.208	0.978	0.213	42	0.669	0.743	0.900	72	0.951	0.309	3.078
13	0.225	0.974	0.231	43	0.682	0.731	0.933	73	0.956	0.292	3.271
14	0.242	0.970	0.249	44	0.695	0.719	0.966	74	0.961	0.276	3.487
15	0.259	0.966	0.268	45	0.707	0.707	1.000	75	0.966	0.259	3.732
16	0.276	0.961	0.287	46	0.719	0.695	1.036	76	0.970	0.242	4.011
17	0.292	0.956	0.306	47	0.731	0.682	1.072	77	0.974	0.225	4.331
18	0.309	0.951	0.325	48	0.743	0.669	1.111	78	0.978	0.208	4.705
19	0.326	0.946	0.344	49	0.755	0.656	1.150	79	0.982	0.191	5.145
20	0.342	0.940	0.364	50	0.766	0.643	1.192	80	0.985	0.174	5.671
21	0.358	0.934	0.384	51	0.777	0.629	1.235	81	0.988	0.156	6.314
22	0.375	0.927	0.404	52	0.788	0.616	1.280	82	0.990	0.139	7.115
23	0.391	0.921	0.424	53	0.799	0.602	1.327	83	0.993	0.122	8.144
24	0.407	0.914	0.445	54	0.809	0.588	1.376	84	0.995	0.105	9.514
25	0.423	0.906	0.466	55	0.819	0.574	1.428	85	0.996	0.087	11.430
26	0.438	0.899	0.488	56	0.829	0.559	1.483	86	0.998	0.070	14.301
27	0.454	0.891	0.510	57	0.839	0.545	1.540	87	0.999	0.052	19.081
28	0.469	0.883	0.532	58	0.848	0.530	1.600	88	0.999	0.035	28.636
29	0.485	0.875	0.554	59	0.857	0.515	1.664	89	1.000	0.017	57.290
								90	1.000	0.000	∞

Index

Hooke's law 141

I

inertia
  principle of 64
integral 93

K

kilo- (metric prefix) 23
kilogram **25**
kinematics 53

L

Laplace 17
Leibnitz 70
light **18**

M

magnitude of a vector
  defined 150
matter **18**
mega- (metric prefix) 23
meter (metric unit) **24**
metric prefixes. *See* metric system: prefixes
metric system **22**
  prefixes 23
micro- (metric prefix) 23
microwaves 18
milli- (metric prefix) 23
model
  scientific 133
models 56
motion
  rigid-body 54
  types of 54
Muybridge, Eadweard 159

N

nano- (metric prefix) 23
Newton
  first law of motion 102
  second law of motion 106
Newton, Isaac 22
  definition of time 25
Newton's laws of motion
  in three dimensions 120
Newton's third law 126

O

operational definitions 24
order-of-magnitude estimates 47

P

parabola

motion of projectile on 119
Pauli exclusion principle 19
period
  of uniform circular motion 177
photon 129
physics **17**
POFOSTITO 128
Pope 37
prefixes, metric. *See* metric system: prefixes
projectiles 119
pulley 142

R

radial component
  defined 179
radio waves 18
reductionism **20**
Renaissance 15
rotation 54

S

salamanders 44
sans culottides 24
scalar
  defined 150
scaling 37
  applied to biology 44
scientific method **15**
second (unit) **24**
significant figures 30
simple machine
  defined 142
slam dunk 56
Stanford, Leland 159
strain 140
Swift, Jonathan 37

T

time
  duration 57
  point in 57
transmission of forces 139

U

unit vectors 156
units, conversion of 28

V

vector 53
  acceleration 161
  addition 150
  defined 150
  force 164
  magnitude of 150
  velocity 160

velocity
    addition of velocities 67
    as a vector 160
    definition 61
    negative 68
vertebra 46
volume
    operational definition 35
    scaling of 37

W

weight force
    defined 101
weightlessness
    biological effects 91

X

x-rays 18

Y

Young's modulus 146

Photo Credits

All photographs are by Benjamin Crowell, except as noted below.

Cover
Moon: Loewy and Puiseux, 1894.

Chapter 1
Mars Climate Orbiter: NASA/JPL/Caltech.   *Red blood cell:* C. Magowan et al.

Chapter 2
High jumper: Dunia Young.   *Rocket sled:* U.S. Air Force.

Chapter 3
X-33 art: NASA. *Astronauts and International Space Station:* NASA.
Gravity map: Data from US Navy Geosat and European Space Agency ERS-1 satellites, analyzed by David Sandwell and Walter Smith.

Chapter 4
Isaac Newton: Painting by Sir Godfrey Kneller, National Portrait Gallery, London.

Chapter 5
Space shuttle launch: NASA.

Chapter 6
The Ring Toss: Clarence White, ca. 1903.

Chapter 7
Aerial photo of Mondavi vineyards: NASA.

Chapter 8
Galloping horse: Eadweard Muybridge, 1878.

Chapter 10

Pluto and Charon: Hubble Space Telescope image, STScI. Not copyrighted.   *Uranus:* Voyager 2 team, NASA. Not copyrighted.   *Earth:* Apollo 11 photograph. Not copyrighted.   *WMAP:* NASA. Not copyrighted.